NEW YORK CITY FARMER & FEAST

Harvesting Local Bounty

EMILY BROOKS

Happy Feasting!

Emily Brooks

BEA '12

gpp®

GUILFORD, CONNECTICUT

All recipes developed by Emily Brooks except for Zucchini Salad with Ricotta on Grilled Bread and Oyster Tacos, both courtesy of Chef Sisha Ortúzar and Riverpark Farm

Photos © Emily Brooks with the exception of the photos on pages 44–47 courtesy of the Horticultural Society of New York, the photo on page 29 by Thinkstock.com, and the photos licensed by Shutterstock.com on pages 11, 15, 19–20, 30, 37, 41, 43, 48–49, 53, 60, 69 right, 83–84, 87, 95–96, 101–102, 107–108, 112, 113 top, 128–129, 134, 139, 147, 151, 155, 163–164, 169, 174–175, 181, 188, 192 left, 193, 198–99, 204, 209, 214–215, 220, 222–223, and 228

Photo of Ms. Brooks © Walter Kidd
Maps: Design Maps Inc. © Morris Book Publishing, LLC

Editor: Amy Lyons
Project Editor: Tracee Williams
Cover Design: Diana Nuhn
Text Design: Sheryl P. Kober
Layout Artist: Melissa Evarts
Photo Research: Lauren Brancato

Library of Congress Cataloging-in-Publication Data

Brooks, Emily, 1975-
New York City farmer & feast : harvesting local bounty / Emily Brooks.
— 1st ed.
p. cm. — (Farmer & feast)
Summary: "Think there are no farms in or around Manhattan? Think again! Urban agriculture and farms are exploding throughout the New York City metropolitan area—from farms to green food-growing rooftops to hydroponics. New York City Farmer & Feast introduces New Yorkers to the surprising bounty of the area and serves as a memento of food experiences for visitors. Profiled are the farmers and produces who feed New York's farmers markets and service top restaurants with locally grown food. Above all, New York City Farmer & Feast is a guide, a cookbook, a reference, and a friendly introduction for anyone who wants to put a face to their local food—and understand where and how its produced"— Provided by publisher.
ISBN 978-0-7627-7951-2 (pbk.)
1. Cooking (Natural foods)—New York (State)—New York. 2. Cooking—New York (State)—New York. 3. Farms, Small—New York (State)—New York. 4. Sustainable agriculture—New York (State)—New York. I. Title. II. Title: New York City farmer and feast.
TX741.B757 2012
641.3'02097471—dc23
2012003217

Printed in the United States of America

10 9 8 7 6 5 4 3 2 1

Farms stands and farmer's markets often come and go, and their produce selection is ever-changing. We recommend you call ahead to obtain current information before visiting any of the locations in this book.

To the fallen and survivors of Hurricane Irene and Tropical Storm Lee, whose unwelcome visits throughout the writing of this book wrought havoc and disaster, leaving lasting scars across all of the Northeast's agricultural landscape.

And specifically to Cheryl Rogowski and the W. Rogowski Farm in Orange County: May those magic beans always grow proudly in your fields. May ten thousand helping hands emerge as you rebuild your entire farm. You are not alone and we need you.

Contents

Preface

Collectively Dirty

Food is one of the most potent yokes that tether all of humankind together. Regardless of race, ethnicity, socioeconomic background, eye color, or demographic location, we all need to eat. We share this human bond with the seminomadic Masai in northern Tanzania and the Inuit people of northern Alaska—most of whom we've never met and know little about. The growth, creation, modification, communal feasting, and act of the giving of food is an art, struggle, and triumph shared by all peoples everywhere—whether one is counting pennies at a bodega, splurging at a five-star restaurant, serving a casserole at a family meal, or foraging through the trash in search of something to eat.

Food is the great equalizer that negates what might be fashionable, how many dollars one might have, what car one might own, what neighborhood one lives in, and what job title (white or blue collar) one might have. The act of eating homogenizes each of us as basically and similarly human.

As humans, we use food as one of our most powerful tools. We use food to define our specific cultural niches. The use of food and feasting defines the basic ritual tenets of all religions—from the taking of communion to participating in Ramadan to consuming only kosher foods. We feast together to celebrate our cultures on the Day of the Dead, Diwali Festival of Lights, and the conventional American Thanksgiving.

We use food as a reward. Cleaning one's room or receiving a vaccination or being someone's significant other might elicit cookies or lollipops or boxes of chocolates. We use food to punish. Misbehaving might get a person sent from the table without dinner. We use food to love and nurture each other. We make chicken soup if someone isn't feeling well and bring food to our neighbors during times of tragedy or death. We use food to differentiate ourselves and create identities. We might label ourselves as vegetarian or vegan or locavore or raw foodist. We use food to elicit or retain some sense of self-control, and, since a child or teenager's only real venue of complete control is food, dinner-table battles ensue over insufficiently clean plates or failure to try vegetables or even to eat at all. We use food as our most powerful global and political leverage via trade tariffs and embargos and International Monetary Fund–World Bank policies.

As a community collective, New York City communally feasts. The daily ritual of breaking bread weds New Yorkers from the deep trenches of crumbling "Food Deserts" to the elitist high-star dining scene, marrying the urban shadeless concrete neighborhoods to winding suburban fertile pastures.

As they feast together, so, too, do New Yorkers farm together. Russian descendants plow their fields alongside African-American teenagers harvesting rows of tomatoes behind fences. We find beehives on the tenth floor above aquaponic fish tanks nestled safely in basements. High rooftop gardens overlook community plots at busy

intersections. Asian schoolchildren are making kale salad on field trips, while tufts of Indian herbs are patchworked into spots of soil between crowded buildings in remote neighborhoods easily forgotten off the beaten path.

Thousands of hands are collectively dirtying themselves deep in the soil. Urban farming in New York City has exploded and will continue to do so exponentially. Backyard gardeners are not only increasing in number but uniting together as food-producing coalitions. Gardens are being planted at every school. Vacant lots will continue to be cleaned, converted, and planted. More rooftops are greening, producing food, collecting rainwater, and producing clean oxygen. More windows are hosting vertical hydroponic gardening systems. More communities are creating their own specialized food cultures and centers and networks.

More hands are getting dirty in the soil.

This collective return to self-supporting agriculture is fueled by thousands of passionate organizations, volunteers, and community programs. More New Yorkers will desire to feed themselves, propelled forward by an innate humanity, an evolutionary desire for physical interaction with Mother Nature, and a social and spiritual hunger for the human connection through food that binds all human peoples and cultures together.

The success of urban farming and local food feasting is being hoisted as a hopeful pennon to the future.

This collective dirt—under fingernails of all NYC age groups, demographics, religions, cultures, and socioeconomic strata—is a powerful freedom.

Get dirty.

And should you seek an autograph, I hope that it is also from every farmer in this book. All but two of the recipes in this book have been developed by me in my kitchen, and although I created them, I have the great pleasure to name these recipes in honor of the New York City farmers that have been highlighted. You'll also notice that most recipes that call for stock do so in a way that does not indicate which type of stock to use. I've left that decision up to you and have provided three stock recipes in the book for you to choose from:

Hemlock Hill Beef Stock (see p. 215)
Mohlenhoff's Turkey Stock (see p. 192)
Rogowski Ruby Red Vegetable Stock (see p. 120)

From their palette to your palate, meet your food and feast well.

Emily Brooks

Introduction

THE URBAN FARMING REVOLUTION

The first and most prevalent reaction I received when telling people that my second book would be *New York City Farmer & Feast* was blank stares and mumblings of complete incomprehension. "Yeah, but all of that food has to be brought in from somewhere else, right? Like upstate New York?" The notion that there is a significant quantity of food grown directly in New York City and its surrounding suburbia is astonishing. The fact that urban farming throughout NYC is tripling and quadrupling at exponential rates is astounding. Nine of the twenty-three counties that make up the greater New York City metropolitan area are surveyed in this book: New York (Manhattan), Kings (Brooklyn), Queens, Bronx, Richmond (Staten Island), Rockland, Orange, Westchester, and Putnam.

Local Food Is the Best Homeland Security

Approximately 80 percent of the US population lives in metropolitan areas. This is in sharp contrast to one hundred years ago, when 50 percent of Americans lived on farms or in small rural communities where they had the knowledge, infrastructure, and available space to produce food and self-sufficiently feed themselves. We no longer have the options and opportunities to rely on community or regional food networks, and today more food is shipped from markets outside the United States than at any time in history. This forced dependence on external food sources makes urban dwellers, or, more aptly stated, 80 percent of the US population, at an extremely high food-security risk.

In 1943, twenty million households (or three-fifths of the entire US population) grew more than 40 percent of all of the vegetables we consumed. The pendulum is starting to swing backward again toward more people participating directly in generating their own food supply. In 2009, forty-three million households, regardless of being urban or rural, were planning to produce some of their own food, up 19 percent from the year before.

The astounding ethnic diversity of New York City is one of the many celebrated reasons why this metropolis is touted as "the greatest city in the world." For the first time, black, Hispanic, and Asian residents of New York City and its suburbs are a majority of the metropolitan area's more than nineteen million residents.

Yet this triumphant increase in the city's cultural diversity has also coincided with a greater than 41 percent decrease in neighborhood supermarkets and a 32 percent increase in pricey and exclusive high-end import food boutiques in more affluent neighborhoods. It is quite probable that

these facts are directly correlated. The number of supermarkets in the lowest-income neighborhoods was almost 30 percent less than the number in the highest-income neighborhoods. Those with lower annual incomes have less discretionary income per month and therefore purchase less food overall. Supermarkets seek to operate in areas with specific, quantifiable characteristics—such as universal food likes and dislikes, and higher income levels, and higher education levels. Lower income, multiethnic neighborhoods are generally less appealing to grocers. As a consequence, the decrease in hundreds of supermarkets makes it extremely difficult and, in many cases, purely impossible for millions of New Yorkers to find fresh and affordable food within walking distance of their homes. Studies by the Department of City Planning estimate that as many as three million New Yorkers live in what are considered "high-needs neighborhoods and communities," which are characterized as having an insufficient quantity of supermarkets selling fresh fruits and vegetables and a subsequently high incidence of health problems. These places are called Food Deserts. The lack of easily available fresh food has prompted city and state officials to convene several task forces to address the public health implications in these lower-income areas. Living

in a Food Desert forces individuals to purchase their food at discount stores or pharmacies or gas stations where there isn't anything fresh, wholesome, or nutritional in sight. Additionally, prices at small neighborhood corner stores are significantly higher than at grocery stores, forcing residents to spend more of their income to receive both less quantity and lower quality foods. It is not surprising that the lack of supermarkets and high rates of preventable illnesses, such as diabetes, are statistics that are directly correlated. This situation is perhaps one of the greatest and most unacknowledged human rights issues we currently face in the United States.

In addition, in 2008 New York spent $6.1 billion dollars attempting to stem the rise in obesity-related health problems and to promote the consumption of fresh fruits and vegetables. One out of every four New Yorkers under the age of eighteen is categorized as obese, and an extreme majority of these kids reside in Food Deserts where they do not have access to fresh and healthy foods. Yet, in New York City, where real estate is at a premium, there remain six hundred stalled construction sites and 596 acres of vacant public land sitting idle, not to mention vacant private land and thousands of acres of rooftops that could be used for urban agricultural production.

A re-localized food system through urban farming thus offers city dwellers sustainable, non-grocery-store alternative access to healthy, affordable food. Too, in shortening the fresh-food-to-table supply chain by expanding urban and suburban regional farming capacities, New York's communities have the potential to reap a long list of benefits, such as increased food security; green space provided by urban farms and gardens; more fresh, wholesome foods and job opportunities where they're needed most; less agricultural pollution and food waste; less traffic and tractor-trailer food transports from states thousands of miles away; and reinvigorated local economies. In short, urban farming is more than an adorable pastime for gardening enthusiasts. It is a vitally important new jobs sector that can play a major role in NYC's health, strength, security, and long-term stability.

For these reasons, we must reestablish the relationship between legislative policy, agrarian culture, and the financial economics of our food supply. In *New York City Farmer & Feast,* you will meet some amazingly ingenious people who ARE using idle land for food production. Their stories are inspiring.

The Power of Community Food-Growing Space

There are well over six hundred community gardens throughout New York City. These community gathering places unite multiple cultures, languages, backgrounds, and socioeconomic demography in green communion. More than 80 percent of the community gardens grow specialized food that is consumed directly by that specific neighborhood. More than 43 percent of these gardens are currently partnered with at least one local school, and another 39 percent desperately seek to do so.

Tracing the history of community gardens in New York City takes us back to the early 1970s when more than ten thousand city-owned vacant lots patchworked across the city as scarred, blackened, fetid trash-dump reminders of poverty or

failure or abandonment or greed. It was the residents of these communities that banded together to erase the yoke of such negative urban blight, beautify their neighborhood, grow and share and commune over food, and feast together.

In 1998, there were nearly nine hundred community gardens throughout NYC, but as the city emerged from its fiscal crisis and the housing development boom began, these carefully tended green oases quickly became prime targets for real estate development. Despite protests, press coverage, and court hearings, many gardens were destroyed. These devastating losses launched some of the most aggressive community-garden advocacy and policy groups in the nation. From the time of disruptive protest and brutal arrests, to the hard and fast battles over garden permanency in both 2002 and 2010, we fast-forward to today—a time when many of these activists and their groups now work amicably side by side with borough presidents, mayors, and councilmen and

Wanaqua Family Farm

councilwomen to shape the policy of NYC's greening urban food production system.

In Manhattan alone, borough president Scott Stringer led the creation of FoodNYC and established food production as an economic stimulus priority by increasing access to farmable land, creating a citywide urban agriculture program, ensuring the permanence of community gardens, and facilitating and developing rooftop farms and greenhouses. In addition, Mr. Stringer went even further and issued public policy mandates for the creation of local food delivery systems, wholesale markets, and processing and distribution facilities; the protection of existing farmland; the increase of the number of farm markets throughout Manhattan; and the implementation of food curriculum in all of their public schools.

The numbers of community gardens are once again on the rise as public policies, such as Manhattan's, are now in place to ensure long-term sustainability of new urban agriculture projects. More importantly community gardens have won the public battle, becoming part of the daily lexicon and the very fabric of New York City as vital, distinguishing marks of social justice. No two community gardens are the same, and they stand out as individualized flags of micro-regional personage and identity.

Sustainable Urban Farming

Growing food in the city isn't the province of privileged youth or aging, Faulkner-reading hippies. Urban agriculture has solid sociological roots and is continuing to explode in decidedly unhip, economically marginal, multiracial neighborhoods. Farming is a social justice movement

that is sweeping throughout every sector of NYC's multifaceted and multiregional populace.

It is vitally important for our economic future to nurture this urban farming renaissance and create long-lasting sustainable city food systems. More elected officials need to declare their commitment to increasing the production and availability of locally grown sustainable food, foster the success of multijurisdictional food-policy councils, purchase local food for official governmental events, divert economic stimulus money into the local food movement, and revise the permitting and zoning processes to make farming in open space easier. In 2009, New York State received $34.8 billion dollars in American Recovery and Reinvestment Act moneys. Except for money diverted into various potential granting programs, there are no specific budget line items to allow any of economic stimulus money to be directly invested in New York's sustainable, self-sufficient, urbanized food systems. Can you imagine the positive impact if this were to change?

Sustainable urban food systems promote health, environmental sustainability, and social responsibility. Localized food grown directly in cities and throughout individual neighborhoods helps to eliminate hunger and ensure access to healthy, nutritious food for all residents, regardless of economic means. Urban farming policies reduce the environmental impact associated with gigantic agribusiness and the centralized food production, distribution, consumption, and disposal while increasing continuous fresh and healthy options in Food Deserts.

The urban and suburban farmers of New York City are charting the future. Throughout *New York City Farmer & Feast,* you will see that

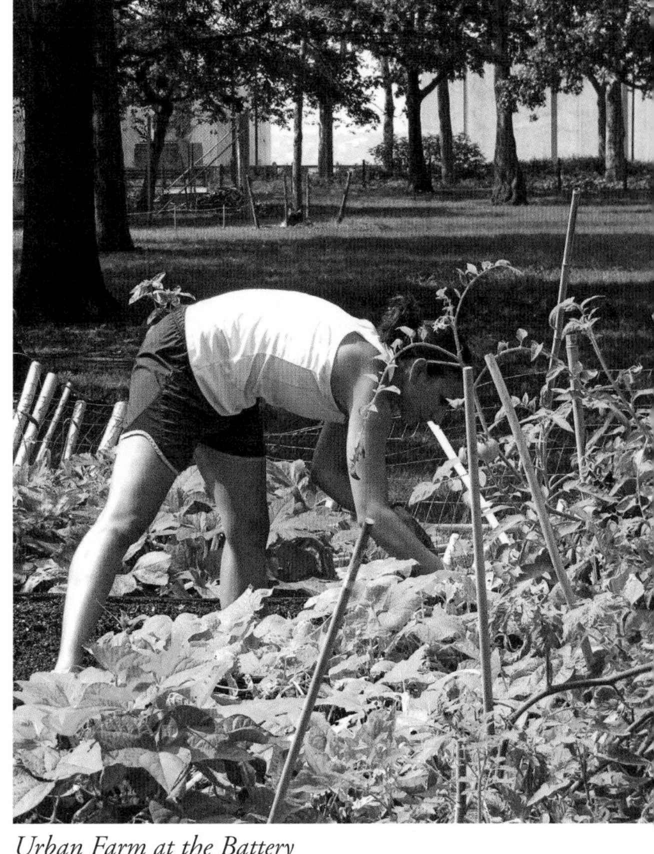

Urban Farm at the Battery

planting a single seed results in much more than a single plant. The positive social impact that regional food production has on its communities is inspirational. Navigating the labyrinth rows of vegetables is revitalizing whole neighborhoods. The economic stimulus and revitalization of New York City could be as simple as a bowl of locally grown salad greens.

All of the farms and farmers featured in the book are open to the public or sell their wares throughout the many NYC farmers' markets. To learn more about these farmers and watch video interviews of their farms, please visit: www.nycfarmerandfeast.com

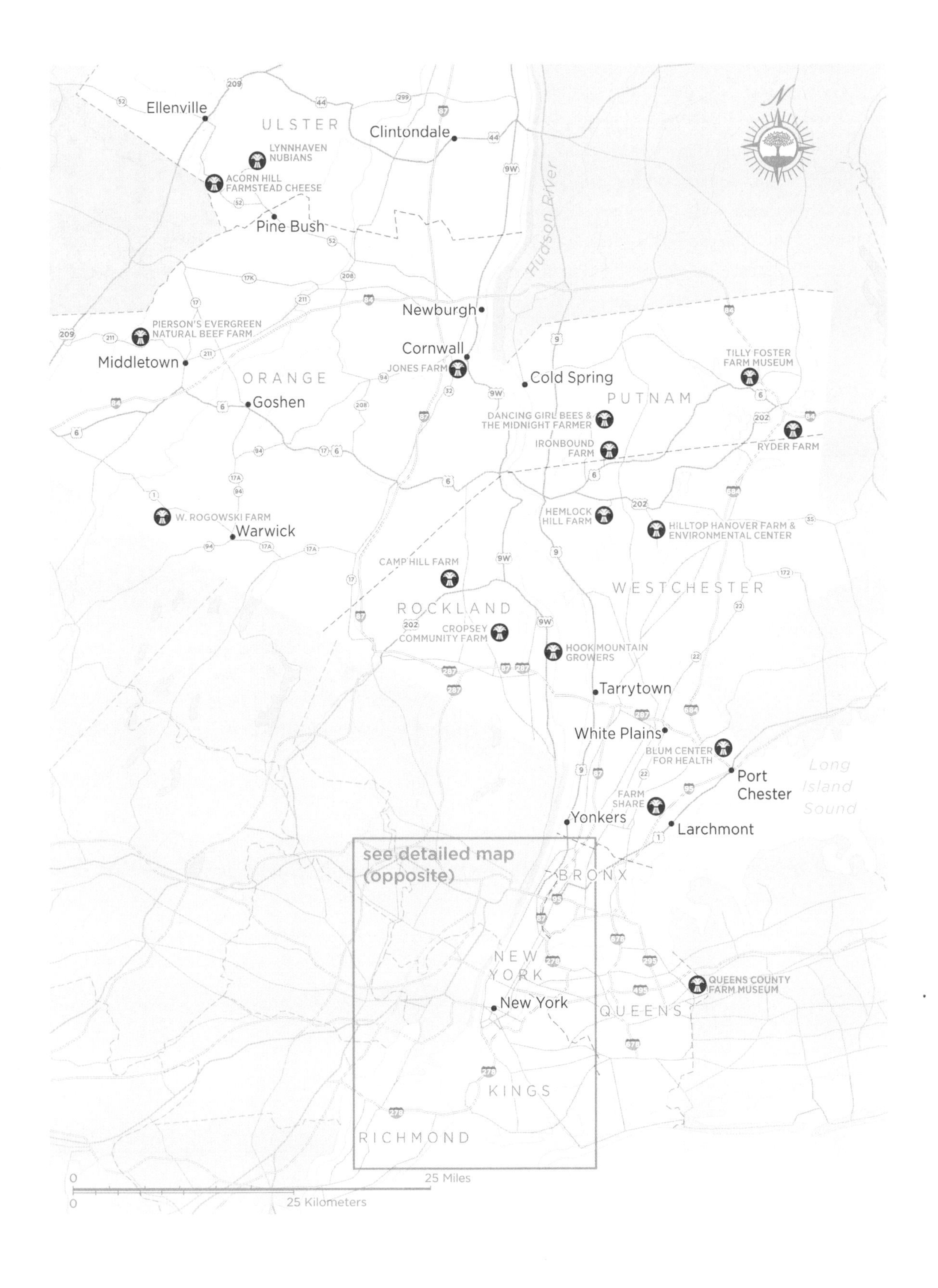
Ellenville
ULSTER
Clintondale
LYNNHAVEN NUBIANS
ACORN HILL FARMSTEAD CHEESE
Pine Bush
Hudson River
Newburgh
PIERSON'S EVERGREEN NATURAL BEEF FARM
Middletown
Cornwall
JONES FARM
ORANGE
Cold Spring
Goshen
TILLY FOSTER FARM MUSEUM
PUTNAM
DANCING GIRL BEES & THE MIDNIGHT FARMER
IRONBOUND FARM
RYDER FARM
HEMLOCK HILL FARM
W. ROGOWSKI FARM
Warwick
HILLTOP HANOVER FARM & ENVIRONMENTAL CENTER
CAMP HILL FARM
WESTCHESTER
ROCKLAND
CROPSEY COMMUNITY FARM
HOOK MOUNTAIN GROWERS
Tarrytown
White Plains
BLUM CENTER FOR HEALTH
Port Chester
Long Island Sound
FARM SHARE
Yonkers
Larchmont
see detailed map (opposite)
BRONX
NEW YORK
New York
QUEENS COUNTY FARM MUSEUM
QUEENS
KINGS
RICHMOND
0
25 Miles
0
25 Kilometers

Washington Heights
BRONX
Hudson River
LA FINCA DEL SUR
BRONX BEES
Harlem
WANAQUA FAMILY FARM
5-BORO GREEN ROOF GARDEN
East River
RIKERS ISLAND GREENHOUSE
RANDALL'S ISLAND CHILDREN'S LEARNING GARDEN
Randalls Island
Rikers Island
College Point
NEW YORK
TWO COVES COMMUNITY GARDEN
Astoria
Manhattan
Jackson Heights
Corona
Long Island City
DICKSON'S FARMSTAND MEATS
RIVERPARK "POP-UP" FARM
LAS DELICIAS PATISSERIE
EAGLE STREET ROOFTOP FARM
Woodside
BERKSHIRE BERRIES
GOTHAM GREENS
FIFTH STREET FARM PROJECT
East River
Greenpoint
Maspeth
QUEENS
URBAN FARM AT THE BATTERY
Williamsburg
BUSHWICK CAMPUS FARM
Woodhaven
Bedford-Stuyvesant
Gowanus
BED-STUY FARM & THE BROOKLYN RESCUE MISSION
East New York
Upper New York Bay
EAST NEW YORK FARMS!
KINGS
Brooklyn
Starrett City
St. George
Flatbush
Canarsie
Port Richmond
JOE HOLZKA COMMUNITY GARDEN
FISH FARM AT BROOKLYN COLLEGE
Boro Park
Rosebank
Jamaica Bay
GREENBELT NATIVE PLANT CENTER
RICHMOND
Bensonhurst
Gravesend
New Springville
GREENBELT CONSERVANCY
Fresh Kills
DECKER FARM
Brighton Beach
Coney Island
Staten Island
Lower New York Bay
Belle Harbor
0
5 Miles
0
5 Kilometers

Chapter 1

Brooklyn (Kings County)

Bed-Stuy Farm & the Brooklyn Rescue Mission

255 Bainbridge St., Brooklyn, NY 11233
(718) 363-3085 | www.brooklynrescuemission.org

Women of the cloth are few and far between; even rarer is one that is stunningly beautiful. Reverend DeVanie Jackson of the Brooklyn Rescue Mission is an inimitable shepherdess for her within-walking-distance denomination in the Bed-Stuy neighborhood of central Brooklyn.

Under DeVanie's direction, the Brooklyn Rescue Mission is an antihunger and food justice program with four main components: the Bed-Stuy Farm, the Malcolm X Boulevard Community Farmers' Market, a food educational program that includes nutrition education and cooking classes, and a food justice policy program that advocates for food rights. The UN Special Rapporteur defines the right to adequate food, or "food rights" as

> a human right, inherent in all people, to have regular, permanent and unrestricted access, either directly or by means of financial purchases, to quantitatively and qualitatively adequate and sufficient food corresponding to the cultural traditions of people to which the consumer belongs, and which ensures a physical and mental, individual and collective fulfilling and dignified life.

The Bed-Stuy Farm is both a food-producing machine, complete with chickens and two different aquaponic fish systems, and a regional outreach agricultural training center in partnership with Will Allen of Growing Power in Milwaukee, Wisconsin, who advocates for food justice and year-round growing in unheated hoop houses made of flexible piping wrapped with plastic to form a roof.

The day we spoke, I found the Reverend packing up food baskets to be distributed to

Reverend DeVanie Jackson

senior citizens through their urban harvest food pantry program. With grant moneys, the mission purchases eggplant, spinach, grapes, and staples such as oatmeal, ravioli, and juice, and distributes them, without discrimination, to elder members of the community. To better serve the needs of her lower-income neighborhood, the lovely Reverend DeVanie Jackson also became a farmer. "Dignity," says DeVanie, "comes from the soil and touching the ground and growing one's own food."

Food justice is simply the right of everyone to have access to fresh healthy food regardless of color, class, or community. The central Brooklyn area is considered a Food Desert with very serious food accessibility problems and very few supermarkets and food outlets. I asked Reverend Jackson: Is the biggest challenge that the fresh and healthy food in the Bed-Stuy neighborhood is too limited or that appropriate quantities of fresh foods are available but priced incorrectly? "The biggest problem is that fresh and healthy food is completely absent," the Reverend says. "You can't find it if you wanted it. No one should have to walk more than four to six blocks to go grocery shopping." Good point. Who can walk a mile or more with a week's worth of groceries strapped in plastic bags over their arms in the blazing heat, driving rain, or slippery snow?

Yet, why is food so absent? "It's part of the community planning process. Large supermarkets have to want to come in and invest in communities because, regardless of the economic status, people still want and need to buy food." Yet, supermarkets continue to stay away. Until this situation changes, the Brooklyn Rescue Mission is bridging the gap.

Having lived through and experienced all sides of this food-availability issue, the senior citizens in DeVanie's flock are her biggest advocates for long-term sustainable community change. Many of these seniors are children of the Great Depression who grew up in the South or the Caribbean, and they know what it's like to forage for fresh eggs and healthy food. Today, they help guide and teach the younger generation who have grown up surrounded solely by city concrete and

can't always tell the difference between a tomato and an apple. Apples are round and red, and A is for Apple. Tomatoes are round and red. A is for Apple. And therefore, round and red tomatoes are obviously apples.

There is absolutely no other program like DeVanie's in the entire United States. I quipped that she is not only the shepherd of her flock, but the farmer for her flock. "They're inherently the exact same thing," she says. "Jesus spent a lot of time talking to farmers, and most of the biblical analogies are food based, such as 'sowing seeds' and 'withering on the vine' and 'harvest time.' I thought I could read the Bible and understand all of those things. I didn't understand the half of it until I became a farmer myself. For example, what in the world does 'the land of milk and honey' mean to a born-and-bred city person who rides the subway?"

To love thy neighbor as thyself is to feed thy neighbor as thyself. The Reverend DeVanie Jackson lives her life practicing the parable of the loaves and fishes, ministering to her community, one bag of grapes at a time.

> Hear this, you elders; listen, all who live in the land. . . . Tell it to your children and let your children tell it to their children, and their children to the next generation. . . . The vine is dried up and the fig tree is withered; the pomegranate, the palm and the apple tree—all the trees of the field—are dried up. And so too then, the joy of mankind has withered away. Joel 1:2–3, 12

DeVanie's Devine Mashed Potatoes

DeVanie and I were chuckling about foods we ate as children that continue to serve as comfort food standbys in adulthood. This recipe is for you, DeVanie. Later in the book, I include a healthier version of Macaroni & Cheese for me.

SERVES 6

Garlic Puree

1 cup peeled garlic cloves
1 cup extra virgin olive oil

Mashed Potatoes

3 pounds red potatoes
2 tablespoons sea salt
½ cup unsalted butter, melted
¼ cup heavy cream, at room temperature
1 cup crumbled soft chèvre goat cheese
¼ cup roasted garlic puree
Sea salt and pepper to taste

To make roasted garlic puree, place 1 cup of peeled garlic cloves in a small ovenproof container. Cover the garlic cloves with the extra virgin olive oil and roast, covered, at 350°F for 1 hour or until garlic is soft and golden. Strain the oil from the garlic and reserve for cooking later. Mash garlic with a fork or with a food processor.

Cut the potatoes into even-size chunks, leaving skins on, and place in cold water to cover with the sea salt. Bring the water to a boil and boil the potatoes until tender, about 15 to 20 minutes.

In a separate bowl whisk melted butter, cream, goat cheese, and roasted garlic puree until smooth; keep warm.

Mash the potatoes with a potato masher or whisk. Add the warmed butter and cream mixture to the hot mashed potatoes and stir gently to combine.

Season to taste with sea salt and pepper and serve warm.

Brooklyn Rescue Mission's Savory Fig Manna

The fig tree at the Bed-Stuy Farm stands is impressive in size, towering into the sky over the entire garden. This recipe, while a little labor-intensive, is worth every minute of preparation time. Many of the world's ills can be soothed with hot manna, friends, spoons, and a good pot of coffee.

SERVES 6

1 large lemon, washed
¼ cup whole raw almonds
3 tablespoons packed dark brown sugar
½ teaspoon sea salt
2 teaspoons fennel seeds
3 tablespoons vodka or dry white wine
1 recipe Zan's Honeyed Tart Crust (see p. 36)
1 large whole egg
1 egg yolk
1½ tablespoons unsalted butter, at room temperature
1 pound fresh black figs, about 20, stems removed and halved lengthwise
2 tablespoons maple syrup
Crème fraîche, clotted cream, or sweetened whipped cream

The day before serving, prepare the lemon and the almonds. Using a mandolin or other slicing equipment, slice the lemon into paper-thin slices. Remove and discard all of the seeds and put the slices in a nonmetallic bowl. Toss the lemon slices with brown sugar, sea salt, fennel seeds, and vodka. Cover with plastic and let stand for 24 hours at room temperature. In a separate bowl, soak the whole almonds in enough water to cover. Place, covered with plastic wrap, in the refrigerator for 24 hours.

Bake Zan's Honeyed Tart Crust as directed. Set aside. Reset the oven temperature to 400°F.

Drain the lemon slices with fennel seeds, reserving the liquid. Drain the almonds, discarding the liquid, and dry the nuts thoroughly on paper towels.

In a food processor, puree the nuts until a paste forms. Add the whole egg, egg yolk, butter, and drained lemon slices with fennel seeds and mix until well blended.

Spread the lemon-almond filling over the bottom of the cooled tart shell and cover with the halved figs, cut side up. Bake at 400°F for about 30 minutes or until the custard has set.

Meanwhile, in a small saucepan, bring the reserved lemon juice to a boil. Add the maple syrup and cook, stirring constantly, until the syrup is reduced to 3 tablespoons of glaze.

Remove the tart from the oven and brush with the lemon glaze. Cool to room temperature before removing the sides of the tart pan. Serve at room temperature with crème fraîche or clotted cream. And, yes of course, Reverend, you may have yours with sweetened whipped cream.

Bushwick Campus Farm

400 Irving Ave., Brooklyn, NY 11227
(646) 393-9305 | www.ecostationny.org

Sean-Michael Fleming is the impetus behind Ecostation NY, a nonprofit whose mission is to explore the connection between human and environmental health. Sean-Michael has found the best tool to support his mission is freshly grown urban food.

Sean-Michael Fleming

His newest venture is a partnership with Bushwick Campus, home to Bushwick School for Social Justice, Academy of Environmental Leadership, and the Academy of Urban Planning. Working with Adam Schwartz at the Bushwick Green Team, they've created a working farm, including a hydroponic fence project, and they sell products through farmers' markets and to restaurants while reinvesting all of the profits back into the Bushwick Campus Farm. In 2012, they intend to implement greenhouses for more year-round growing and community involvement.

And, too, they're building their Green Machine. Currently zipping products from the farm to market in rented U-Haul vans, the new Green Machine will be a large step van, fueled by vegetable oil and cooled with solar power. Sean-Michael has a functional-antique artistic style and eclectic flare. No doubt a retro-chic mobile food wagon will catch the delighted attention of most people in the Bushwick neighborhood.

A farmer? An entrepreneur? A social activist? Sean-Michael is actually an artist. A few years ago, he underwent a deep personal journey, and harking back to the days of his youth, he returned to nature as a source of solace. "I looked around in my community and saw an underappreciation of the natural world, which is not surprising in an urban setting," he says. "You can spend your entire day without touching raw, bare earth." Starting his work with community gardens and

then transitioning to urban farming, he has made Ecostation NY a mission-driven embodiment of his personal journey back to Mother Nature.

Travis Tench, the Ecostation NY farm market manager, is looking forward to his new Green Machine. As the Ecostation NY markets continue to grow, offering cooking classes at every market every week, Travis watches the community respond most positively to the combination of fresh food and food education. Accommodating a wide mix of demographics, such as Caribbean descendants, immigrants from South America, seniors, young mothers with children, and the young artists flocking to the neighborhood, the Bushwick Campus Farm responds accordingly by growing a plethora of tomatillos, collard greens, and bitter melon.

Ecostation NY is a solid presence in this Brooklyn community, with a long list of collaborators and supporters, such as New York Cares, GreenThumb, Green Guerillas, the New York State Department of Environmental Conservation, and Earth Matter, just to name a few. Their outreach presentations emphasize all the health benefits of eating farm-fresh vegetables and showcase farmers' markets as an important counterbalance to the growing number of factory farms, while simultaneously teaching people to grow and cook the foods they most enjoy.

I asked Travis how he liked bitter melon. Chuckling, he admitted, "I'm still learning," he says. "I guess we all are, together."

Bushwick Campus Lamb & Fig Curry

While the particular version of this recipe uses lamb, don't feel limited to that. Feel free to use any protein source such as chicken or beans or beef—or substitute meaty vegetables for the lamb, such as eggplant. If you'd like a milder version, ignore the addition of the jalapeño chile! For medium-hot curry, deseed and devein the jalapeño chile!

SERVES 4–6

2 garlic cloves, minced
1½-inch piece fresh ginger, peeled and minced
2 teaspoons sea salt
1 medium onion, thinly sliced
¼ cup olive oil
1½ pounds boneless leg of lamb, trimmed and cut into ¾-inch pieces
2 teaspoons ground cumin
2 teaspoons ground coriander
1 teaspoon ground turmeric
½ teaspoon fenugreek leaves or ¼ teaspoon ground fenugreek
2 tomatoes, peeled, seeded, and finely chopped
1 teaspoon tomato paste
2 cups stock of your choice (see p. ix) or water
1 tablespoon maple syrup
1 jalapeño chile, stemmed and cut in half through the stem
¼ cup dried figs, chopped coarsely, or 1½ cups sliced fresh figs
4 tablespoons minced fresh cilantro leaves
Sea salt and pepper to taste

Mash garlic, ginger, and sea salt with a mortar and pestle to create a thick paste.

In a large, deep skillet or sauté pan, sauté onion in olive oil over medium-high heat until softened, about 4 minutes.

Stir in the garlic-ginger paste, lamb, ground spices, fenugreek, tomatoes, and tomato paste. Cook, stirring constantly, until the liquid evaporates, the oil turns orange, and the spices begin to fry, about 5 to 7 minutes.

Add the stock, maple syrup, jalapeño, and figs. Bring to a simmer and reduce the heat to medium low. Cover and simmer until the meat is tender, about 40 to 50 minutes. Remove the jalapeño chile, stir in the cilantro, and season to taste with sea salt and pepper. Serve immediately.

Sean-Michael's Green Peas with Allspice

If you're tired of eating fresh peas right out of the pod in the spring (it seldom happens, but just in case), here is some comfort food which goes well with a little dry white wine and a late night movie. This dish is best eaten with your feet up on the furniture.

SERVES 4

3 tablespoons unsalted butter
1 cup minced red onion
1 teaspoon minced garlic
3 cups sliced mushrooms
¼ cup cold water
4 cups fresh green peas
¼ teaspoon sea salt
¼ teaspoon pepper
½ teaspoon ground allspice
Sea salt and pepper to taste
Hot cooked, buttered rice of your choice

Melt the butter in a large saucepan over medium heat, add the onions, and cook, stirring occasionally, until golden and soft. Add the garlic and cook until it just begins to brown. Add the mushrooms and continue sautéing until the mushrooms are deeply golden.

Add the water and peas. Cover and cook over medium heat for about 6 minutes, until the peas are bright green and still have a slight bite. Add the sea salt, pepper, and allspice. Cook uncovered, stirring constantly, for an additional 2 minutes. Season to taste with sea salt and pepper. Serve over piping hot rice with a pat of butter.

Tusha Yakovleva

Eagle Street Rooftop Farm

44 Eagle St., Brooklyn, NY 11222
(917) 656-1161 | www.rooftopfarms.org

On the shoreline of the East River and with sweeping views of the Manhattan skyline, Eagle Street Rooftop Farm is a six-thousand-square-foot organic green-roof garden resting above a nondescript warehouse. These rows and rows of vegetables surrounded by egg-laying chickens, rabbits, and honey, hidden three stories up in the air from street view, are somewhat bizarrely amusing. Not a laughing matter, such is the ingenious beauty and brilliant functionality of urban agriculture.

Unlike shares in community supported agriculture (CSA) organizations in more rural areas that package only what they produce, the Eagle Street Rooftop Farm has a solid and symbiotic partnership with the young Farmer John Ronsani at Lineage Farm in the Hudson Valley to expand both the variety and the quantity of products available to farm-market shoppers, CSA customers, and restaurants.

Young Tusha Yakovleva, who started her career in journalism, helps manage the farm in its third season following in the footsteps of founding members Annie Novak and Ben Flanner. From writing on green topics to becoming a full-fledged farmer herself, Tusha feels that she can contribute more to her community by working in the soil and feeding chickens than by climbing the corporate ladder in Prada heels. She is not alone. While the typical rural farmer is a male in his sixties, urban agriculture is swarming

with young men and women in their thirties who have higher-education degrees and are abandoning their chosen career paths and diving headfirst into food production.

Savvy women like Tusha are taking the lead. Into the aging farming family, they're bringing a global environmental worldview, social media and online digital marketing, and nouveau technology. Tusha stands as an equal in the regional food system, side-by-side with her more-typical farming elders who still harvest hundreds of acres with plows and tractors just north of the city.

Farmer John often gets asked why he chose agriculture as a livelihood. "In learning about food," he says, "I uncovered a rich family history. I heard stories of my grandfather and his brothers and my great grandfather, the 'banana king' of Hudson, who would bring his horse-drawn wagon down to the river to pick up his order and would drop off bushels of bananas to local merchants on his way up to Warren Street. I have found emotional security in continuing the tradition of care that has gone into this land."

New York's regional foodshed is undergoing a colossal metamorphosis with young "farmer pups" joining wizened grandfathers at the same table. A deep sense of history and community still remains the common unifying bond regardless of demographic, digital savvy, or the height of the dirt—whether expansive tillable pastures or carefully constructed mounds atop a three-story warehouse.

Eagle Street Hearty Chickpea Stew

So the fall is coming. Nights are shorter, temperatures are cooler, and yet there is still so much to do! This recipe is a fix-on-Sunday-eat-for-days slow food hug. And, the debate rages about peeling or not peeling carrots for cooking. I say, that if you know your farmer, keep as many skins on your root vegetables as you possibly can! The flavor and extra nutrition are a shame to waste!

SERVES 4

½ pound dried chickpeas
¼ pound dried large white beans
1 pound sweet chorizo, sliced
2 tablespoons olive oil
6 whole carrots, sliced into ½-inch rings
6 cups swiss chard or collard greens cut into 1-inch pieces
1 smoked ham hock
½ teaspoon smoked paprika
1 clove garlic, sliced
2 shallots, sliced
6 cups stock of your choice (see p. ix)
2 roasted red peppers in oil, drained
½ teaspoon whole cumin seeds
¼ teaspoon ground nutmeg
4 peppercorns
Sea salt and pepper to taste
Whole plain yogurt and fresh lemon juice for serving (optional)

Soak the chickpeas and beans overnight in cold water to cover.

In a large stockpot, sauté the chorizo in olive oil until browned. Drain the beans and add to the sausage in the stew pot along with the carrots, swiss chard, ham hock, smoked paprika, garlic, shallots, and stock. Bring to a boil, cover, and simmer on reduced heat for 2 to 2½ hours or until the chickpeas and beans are tender.

In a food processor, add roasted red peppers, cumin seeds, nutmeg, and peppercorns. Puree into a thick paste. Stir into the stew and season to taste with sea salt and pepper.

Remove the ham hock and shred the meat, discarding any bone or fatty gristle. Return shredded meat to the stew. Season to taste with sea salt and pepper. Serve warm garnished with a dollop of whole yogurt and a drizzle of fresh lemon juice if desired.

Tusha's Roasted-Garlic Tomato Salad

This is perfect summer picnic food! Have leftovers? Place the remaining salad in your favorite quiche recipe, or puree the salad in a food processor and use it as a lovely bread spread or as a way to flavor soups and stews.

SERVES 4

1 (16-ounce) can garbanzo beans, drained and rinsed
2 cups seeded, diced tomatoes
¼ cup minced fresh parsley leaves, divided
1 teaspoon sea salt
½ teaspoon pepper
4 unpeeled garlic cloves
1 jalapeño chile, stem removed
¾ cup olive oil
¼ cup balsamic vinegar
Sea salt and pepper to taste

In a large bowl, combine garbanzo beans, tomatoes, and half of the parsley. Sprinkle with sea salt and pepper. Cover and set aside.

Place unpeeled garlic cloves and jalapeño chile in a small skillet over medium heat. Roast, turning frequently, until soft and starting to brown, about 10 minutes for the chile and 15 minutes for the garlic. Cool slightly and then remove the skins from the garlic and roughly chop the chile without discarding the seeds.

In a blender, combine the garlic, chile, oil, and vinegar. Puree until smooth and season to taste with sea salt and pepper.

Pour over the garbanzo beans and tomatoes. Allow salad to marinate for at least 1 to 2 hours. Before serving, add the remaining parsley leaves and toss to combine. Season to taste with sea salt and pepper. Serve chilled or at room temperature.

East New York Farms!

613 New Lots Ave., Brooklyn, NY 11207
(718) 649-7979 | www.eastnewyorkfarms.org

Utilizing a different model than most typical "farms," East New York Farms! is actually a cohesive network of urban farms and gardeners consisting of two official farming plots and more than fifty ad hoc people who utilize their backyards and open community-garden space to produce food.

Catching up with farm manager David Vigil at the East New York Farmers' Market in Brooklyn is a bit like stepping back in time. My mind replayed visions placed there by old films that falsely glorified the lives of African Americans in the South during the 1930s. More than sixty years later, that shiny, happy cinematography is finally a truth in this neighborhood that teems with playful children zipping through the water from fire hydrants, sidewalk barbers, and neighbors chatting contentedly in the shade.

The youth farm in the New Lots neighborhood offers paid agricultural internships to teenagers creating food to sell at the farmers' market. These kids learn not only how to produce food, but they practice maintaining the open community garden and operate the farm stands, retailing food to the public. Using peer education, new interns attend workshops taught by their graduated predecessors. As these kids complete their internships so, too, will they assume the role of teachers as they pass their urban-farming knowledge down to their neighbors two or three grades behind them in school.

The official mission of East New York Farms! is to organize youth and adults to address food insecurity in their community by way of local, sustainable agriculture. And while they are clearly succeeding in their mission, it is obvious that these kids are also absorbing lifelong benefits from this nonprofit endeavor. Through peppers and compost and melon vines, East New York Farms! illustrates an ancient Chinese proverb in daily action: Give a man a fish; you have fed him for today. Teach a man to fish; and you have fed him for a lifetime.

David Vigil

Insidiously hateful is the stereotype that young and idle African-American hands primarily wreak havoc. I was elated to watch David Vigil take this common misperception and clearly demonstrate its caustic fallaciousness. Because of his leadership, hundreds of thirteen- to fifteen-year-old multiethnic hands are consciously rebuilding their own community, one basket of food at a time. And, thousands more are on waiting lists. This year alone, five times more young people applied for the golden intern spots at East New York Farms! than the capacity.

Kareem, Quadel, Paula, Elijah, and Damian, just to name a few, are among our burgeoning new class of heroes. Before they even have their driver's licenses, these kids are standing to support their neighborhoods, diligently and deliberately nourishing them, and changing the very fabric of their own future.

David's New Year's Black-Eyed Peas

Rumor has it that consuming black-eyed peas on New Year's Day brings wealth, success, and riches. These talisman traditions of good luck are recorded in the Babylonian Talmud: "Abaye said, now that you have established that good-luck symbols avail, you should make it a habit to see *rubiya* (black-eyed peas) on your table on the New Year." It is believed, however, that this "good luck" custom is quite accidental and results from translation confusions between Aramaic and Arabic. Correctly translating the Aramaic texts, we should be eating fenugreek on New Year's Day. Oh well.

SERVES 6-8

1 pound dried black-eyed peas
1 cup diced carrots
2 cups diced onions
1 cup finely diced sweet potato
2 cloves garlic
1 orange, peeled, sections and seeds removed, and diced
3 tablespoons fresh thyme leaves
½ teaspoon black pepper
¼ teaspoon cayenne pepper
4 tablespoons olive oil
4 cups stock of your choice (see p. ix), plus additional as needed
1 meaty ½ pound ham bone or ½ pound diced smoked turkey
Sea salt and pepper to taste

Soak the beans overnight in enough cold water to cover. Drain, rinse, and set aside.

In a large stockpot, sauté the carrots, onions, sweet potato, garlic, orange pieces, thyme, black pepper, and cayenne pepper over high heat in olive oil until all the vegetables have sweated their liquid and are starting to brown, about 8 to 10 minutes. Add the stock, ham or turkey, and beans. Bring to a boil.

Reduce heat, and continue to cook, covered, over low heat for about 45 minutes or until the peas are tender but not mushy. Add more stock if the peas become too dry. Remove the ham from the broth and shred coarsely. Return the meat to the beans. Season to taste with sea salt and pepper.

East New York Farms!' Hoppin John

Hoppin' John is the Southern United States' version of the rice and beans dish traditional to West Africa. This "Americanized" black-eyed pea tradition of a New Year's Day good luck feast also calls for the addition of hearty winter greens such as kale or chard because they are as green as the color of money. All of these green vegetable "money reminders" are grown throughout the gardens of East New York Farms! and give a whole new meaning to "Show me the money!"

SERVES 6-8

4 cups cooked David's New Year's Black-Eyed Peas (see p. 19)
2 tablespoons olive oil
¼ teaspoon ground mace or nutmeg
1 teaspoon orange zest
1 cup long grain white rice
2 cups stock of your choice (see p. ix)
Juice of 1 orange
½ teaspoon sea salt
½ teaspoon black pepper
1 cup fresh corn kernels cut from the cob
3 tablespoons unsalted butter
Sea salt and pepper to taste

Cook David's New Year's Black-Eyed Peas using ½ pound smoked turkey instead of ham, if it is an option.

In a large pot with a tight-fitting lid, heat the olive oil and sauté the mace and orange zest for 10 seconds over medium heat. Add the white rice and toss vigorously to coat with oil and toast slightly. Add the stock, orange juice, sea salt, and pepper and cook, covered, over low heat for 6 or 7 minutes or until the rice is almost tender.

Stir in David's New Year's Black-Eyed Peas and the corn. Cover and simmer over low heat until all the liquid is absorbed and the rice and corn are tender. Add unsalted butter and stir to melt. Season to taste with sea salt and pepper and serve warm.

Fish Farm at Brooklyn College

2900 Bedford Ave., Brooklyn, NY 11210
(718) 951-5000 | www.brooklyn.cuny.edu

Martin Schreibman is a distinguished professor emeritus of biology at Brooklyn College. For the last twelve years, Martin has championed urban aquaculture fish farming and water-reuse aquaponic and aquaculture systems. The term *aquaponics* combines aquaculture, the raising of fish in tanks, with hydroponic, the cultivation of plants in water. The beautiful thing about aquaponics is the use and reuse of precious water to grow plants right alongside the growing fish as an integral part of the same shared system.

Deep in the bowels of Ingersoll Hall, a campus building of raging-red brick festooned with towers of lush ivy, Professor Schreibman has been laboriously testing and studying urban fish farming. Now he can create and design specific systems for specific locations. His current

project is implementing an aquaponic fish farm in the backyard of the Brooklyn Rescue Mission.

Martin Schreibman is a research-focused water farmer. "My mama never wanted me to get dirt on my hands," Martin jokes. More seriously, he points out that we have a major global famine issue. In a time where one out of six Americans don't have enough accessible food, growing fish aquaponically (using closed-water recirculated systems without soil, fertilizer, chemicals, and antibiotics) will be the future of food production throughout the entire world. He produces tilapia and other freshwater and saltwater fish, warm- and cold-water fish, and everything in between. And, actually, anyone can follow his example. Quite feasibly, every household could go fishing in their basement.

Martin has a deliciously silly sense of humor that is a bit unexpected from someone with a fancy emeritus title who spends his spare time enjoyably devouring research on the intricate foibles of enzymatic pathways. Yet hanging out with the professor at his glorified fish tanks is a fun and giggly excursion into the center of schools of fish that most of us don't get to experience—even if we take to the high seas on a fishing trawler. As Martin moves swiftly around with his hands in the water and plays with his swimming food, it's obvious his mama may have scolded him about getting dirty but forgot to mention anything about not getting wet.

Most of us only see tilapia as white fillets under plastic wrap. As a chef, I found it amazingly interesting to watch and touch swimming tilapia an arm's length away. Fish and seafood are a food-producing landscape that most of us cannot partake in from beginning to end, but that is changing because of progressive developments in aquaculture. The depths of the oceans and the special equipment required to farm for seafood prevents consumers from truly understanding all things fishy.

It's true that aquaculture has gotten off to a rocky and unhealthy start, and the question of whether to eat or not to eat farm-raised salmon still plagues many health journals to the great dismay of utterly confused consumers. "We've gone through a number of technological farming revolutions," states Martin. "Have we had a few problems? You bet. Is aquaculture where we are now totally without problems? Absolutely not. Is it fully sustainable yet? Absolutely not, not yet. That's what my work dedicates itself to: identifying what the problems are and solving them."

And he does, smiling (and dripping) all the way.

TILAPIA DR

Martin Schreibman

Martin's Creamy Tomatillo Tilapia

Fish recipes tend to be pretty boring—especially those that call for a mild-flavored fish such as tilapia. With that said, you can get away with pretty wild flavor combinations when cooking white fish. Play a bit!

SERVES 4

1½ tablespoons olive oil
4 large tilapia fillets, about ½ pound each
½ teaspoon sea salt
½ teaspoon freshly ground white pepper
¼ cup heavy cream
½ cup Fifth Street Roasted Tomatillo Salsa (see p. 74)
Sea salt and pepper to taste

In a 10-inch nonstick skillet, heat the olive oil until hot, but not smoking. Sear tilapia fillets over medium-high heat until the top becomes opaque, about 5 minutes. Sprinkle with sea salt and pepper. Turn the fillets over and continue cooking on the other side until tilapia begins to brown, about another 3 or 4 minutes depending on the size of each fillet.

Remove tilapia to serving plates and cover to keep warm. To the hot skillet, add the cream and Fifth Street Roasted Tomatillo Salsa. Cook, stirring constantly, until a rich and creamy sauce develops. Season to taste with sea salt and pepper. Spoon over the tilapia fillets and serve immediately.

Brooklyn College's Grilled Radicchio Salad with Shrimp

Radicchio, a leafy chicory, has been noted throughout history as an aid to insomniacs. In the United States, we're a little slow to catch on to the Italians' preference for grilling their chicory and we prefer, instead, to eat it mostly raw in salads. No matter. If your radicchio has its root attached, feel free to dry and grind it up to make a lovely coffee substitute with the added benefit of ridding yourself of intestinal worms (in case you need help in that department). Well, Martin thought that was funny.

SERVES 4

2 teaspoons finely minced garlic
4 tablespoons olive oil, divided
⅓ cup freshly squeezed orange juice, divided
1 teaspoon orange zest

3 teaspoons freshly squeezed lemon juice, divided
2 teaspoons sea salt, divided
2 teaspoons freshly ground black pepper, divided
1¼ pounds large, deshelled and deveined shrimp
2 tablespoons unsalted butter, softened to room temperature
2 tablespoons finely minced shallots
1 tablespoon minced parsley
1 large radicchio, cut lengthwise into 8 wedges with core intact
1 teaspoon balsamic vinegar
2 cups baby salad greens
Sea salt and pepper to taste
¼ cup chopped, toasted walnuts

At least an hour before serving, sauté garlic in 2 tablespoons olive oil in a small saucepan over medium heat until fragrant and beginning to crisp. Add 2 tablespoons orange juice, orange zest, 2 teaspoons lemon juice, 1 teaspoon sea salt, and 1 teaspoon pepper. Whisk briefly to combine and remove from heat. Add the shrimp and marinate for 1 hour, or set aside until ready to marinate the shrimp.

Cream the butter with the shallots, remaining 1 teaspoon lemon juice, parsley, and remaining sea salt and pepper. Set aside.

Brush the wedges of radicchio with the remaining 2 tablespoons olive oil. Grill, turning often, until the surfaces are browned and the interior is tender, about 5 minutes. Place the radicchio in a covered dish and set aside.

In a medium sauté pan, sauté the shrimp and marinade over medium heat, turning often, until they're cooked and the marinade is reduced and syrupy. Add the shallot butter and cook, stirring constantly for 1 more minute. Remove from heat and set aside.

In a separate bowl, whisk the remaining orange juice and balsamic vinegar. Pour into the sauté pan used for the shrimp. Quickly bring to a boil and reduce slightly.

Divide the baby greens on four plates. Add the warm radicchio and shrimp. Drizzle with deglazed orange juice and balsamic vinegar. Season to taste with sea salt and pepper. Sprinkle with toasted walnuts and serve.

Gotham Greens

810 Humboldt St., Brooklyn, NY 11222
(646) 458-1747 | www.gothamgreens.com

Gotham Greens is a marvel of hydroponic food-growing technology. Nearly every element of this coliseum of greenhouse glass is digitalized and fueled by energy captured by solar panels and thermal design features. It is New York City's first commercial-scale greenhouse farm, totaling fifteen thousand square feet.

Jenn Nelkin and Viraj Puri

Gotham Greens' methods of farming yield twenty to thirty times more product per acre than field production while eliminating the dependency on arable land. Knowing that agriculture is the largest consumer of freshwater on the planet, this rooftop farming system uses a recirculating hydroponics system, which uses twenty times less water than conventional agriculture while eliminating all agricultural runoff.

Viraj Puri, one of the cofounders with Eric Haley, understatedly comments that they operate "a state-of-the-art greenhouse." Actually, this is a controlled-environment agricultural phenomenon, complete with a multitude of sensors linked into a mammoth computer system. Based on various atmospheric data and growing requirements programmed into the computer, the digital system then deploys appropriate greenhouse equipment to open roof vents and passive side vents, actuate fans, draw shade curtains, deploy heaters, and control watering levels. This resembles the backyard garden of the Jetsons.

Farmer and greenhouse director Jenn Nelkin, the design wizard and mastermind behind all of Gotham Greens' food production, has worked across the world in extreme food-growing environments. A nationally renowned expert in controlled-environment agriculture, she has grown food in the frozen tundra of Antarctica, the thirsty deserts of Arizona, and on a floating barge right in the Hudson River.

"Controlled-environment agriculture lends itself very well to New York City, where we don't have a lot of arable land or fertile soil," Viraj says. "Rooftops are an underutilized resource, and it is very important to us to grow year-round, bringing the consistency of food supply reliability and product quality." With eight million people living on the land of New York City, this adaptation to the NYC environment is fundamentally important. Compared to other urban farms, Gotham Greens produces an extraordinary amount of food. "We're still really tiny, only a third of an acre," Viraj notes. And yet, they produce weighable tons of greens and herbs above the Greenpoint Wood Exchange building.

Jenn harvests all things green before breakfast so that they can be on our plates by lunchtime. Too, Gotham Greens doesn't just talk blindly about being "local" or "sustainable" or "natural." They are actively practicing what they preach: producing food while also preserving water and soil resources, reducing the transportation burden of product to market, reducing harmful chemical use in food production, ensuring the fair treatment of their workers, and spending as much of their profits as close to home as they are able.

There are many ways to farm responsibly and sustainably. Viraj, Eric, and Jenn's methods are specifically tailored to their unique geographic location in Brooklyn. Their business idea was right on target: In less than two years of operation, the demand for their vegetables has already outstripped their available supply.

When asked if he missed getting dirty in the soil, Viraj smiles. He surveys the phenomenal skyline view and the cozy, bug-free solace inside glass walls echoing with the sounds of bubbling waterfalls that nourish thousands of growing plants.

Gotham Greens' Warm Frisee Salad

Frisee is a member of the chicory family and you can find it in just about any range of colors. As with all leafy vegetables, the darker the color of the leaves the stronger you will find the taste. Frisee is slightly bitter to begin with and this recipe calls for apple cider and dried fruit to sweeten the darker taste of this leafy green. Tired of salad in winter? This has you covered.

SERVES 4-6

1 pound baby red potatoes
2 teaspoons sea salt
1 bay leaf
3 cloves
4 whole black peppercorns
3 tablespoons dried black currants
3 tablespoons apple cider
8 cups frisee lettuce leaves
2 tablespoons white wine
1 teaspoon Dijon mustard
⅓ cup olive oil
2 teaspoons toasted cumin seeds
Sea salt and pepper to taste
4-6 prosciutto pieces, sliced into strips

Place the potatoes, sea salt, bay leaf, cloves, and black peppercorns in a large saucepan. Cover the potatoes with water and bring to a boil over medium heat. Lower the heat and simmer partially covered until cooked, about 20 minutes.

Meanwhile, combine black currants with the apple cider and set aside to soften.

Tear the frisee leaves into bite-size pieces and place in a large salad bowl. In a separate large serving bowl, mix the white wine, Dijon mustard, apple cider, and black currants. Slowly whisk in the olive oil to form a vinaigrette.

Drain the potatoes and discard the bay leaf. While they are still warm, cut the potatoes in half and add to the salad dressing. Immediately add the frisee leaves and toss to combine. Add the toasted cumin seeds and toss again. Season to taste with sea salt and pepper. Add the prosciutto strips, toss one last time, and serve immediately.

Farmer Jenn's Twofold Garlic Escarole Salad

Another winter salad recipe, escarole is a less-bitter type of chicory with broad, pale green leaves. If you travel to Antarctica like Jenn did, then you can eat this salad to warm up! This works beautifully as a side dish to any fall or winter roast or as a topping on grilled hamburgers. My favorite, however, is serving Jenn's salad under a beautifully poached egg. Who says you can't have vegetables for breakfast?

SERVES 4

1 head escarole
6 tablespoons olive oil, divided
½ lemon, sliced thinly crosswise into rings
1 shallot, finely minced
3 cloves garlic, peeled and finely sliced
3 tablespoons freshly squeezed lemon juice, divided
Sea salt and pepper to taste
1½ teaspoons fresh thyme leaves
3 tablespoons chopped fresh flat-leaf parsley

Wash the escarole and cut the entire head of escarole in half horizontally. Shred the top half's tender yellow-green leaves and place in a large salad bowl and set aside. Discard any coarse stems from the dark green leaves of the base end and finely slice remaining leaves into ¼-inch strips. Set aside.

In a large frying pan, heat 2 tablespoons olive oil over medium heat. Add the lemon slices and sauté until browned and fragrant. Remove the lemon slices to a side plate for garnish. To the hot frying pan, add the shallot and cook, stirring until fragrant, about 1 minute. Add the garlic and sauté, stirring constantly, until the garlic begins to brown, about 30 seconds longer.

Pour in 2 tablespoons lemon juice, 2 more tablespoons of the oil, and add the sliced dark green escarole leaves. Sauté until the greens begin to wilt, about 1 minute. Season to taste with sea salt and pepper. Remove the pan from the heat and stir in the last remaining 2 tablespoons of olive oil and fresh thyme leaves. Pour the mixture over the raw escarole leaves in the salad bowl. Add the parsley and toss well. Season to taste with sea salt and pepper.

Serve immediately, garnished with cooked lemon rings.

OOD EXCHANGE
GREE
GMDC
HUMBOLDT ST
MESEROLE AV
ONE WAY
FIVE STAR
CARTING

Chapter 2

Bronx (Bronx County)

Bronx Bees

654 Manida St., Bronx, NY 10474
(917) 880-8159 | www.bronxbees.com

Scaling three stories of an old brick building on an antiquated ladder fire escape is easier than it looks. Zan Asha skittered ahead easily, like a happily chatting, agile monkey, while I tried my best to look graceful behind her. At the pinnacle of our destination were six swarming beehives in the center of her roof. The sky nearly blackened with starving bees released into the bright sun after almost two weeks of solid rains from Hurricane Irene and Tropical Storm Lee. Zan, with her dreads mounded imposingly atop her head, stood smiling with excitement as masses of bees swarmed about her.

Undeniably vivacious and known for occasionally dressing like a gypsy merely for random chuckles, Zan is a third-generation bee farmer. Her Hungarian grandfather, Ference Jogg, deserted

his army post on more than one occasion during World War II. Refusing to fight for the Nazis, he was captured and punished severely for his desertion and held as a prisoner of war by the Hungarian army. Even after the Russians defeated the Nazis, it took Ference seven years to return back home to his wife and their family farm and to meet his seven-year-old daughter for the very first time. Compliments of Joseph Stalin's regime, Zan's grandfather's farm was dismantled and absorbed into the endlessly churning Communist machine.

Together with his three brothers who had survived the brutal atrocities of war, Ference turned to placidly harvesting honey from wild bees. Unlike his granddaughter Zan, who has modern domesticated honeybees in movable-frame, man-made hives, Ference and his brothers used bee skeps, which are baskets made of straw and placed open end down in the wilderness. Since 1998, this beekeeping practice has been outlawed by the United States. To extract the honey from the interior of the basket skeps, the entire colony must be destroyed either by fire or sulfur or by carefully squeezing the entire skep and its colony through squeezing vices.

Lo these many years later, Zan is carrying on her grandfather's family tradition on a rooftop in the Bronx 4,375 miles away from the Hungarian forest.

Zan Asha is not only an organic beekeeper, but the creator of the Bronx Bees Company,

Zan Asha

which is a full-service bee-support organization. Bronx Bees provides unique chemical-free beekeeping maintenance services in the NYC metro area as well as education and training services across the United States. Zan is available to set up and maintain hives, clean hives every spring, and split hives to prevent swarms.

As I slowly inched my way down the ladders, tightly gripping every rung, I fearfully wondered how the bees got up on the roof in the first place and how the honey gets back down. I realized that in meeting Zan, I had just met the real-life version of Disney's moxie Aladdin.

Zan's Honeyed Tart Crust

If you're afraid of baking. don't be. If you'd prefer instead to run to the grocery store to buy pre-made tart or pie crusts, then know that you're stuck with subpar products that have added fillers such as bitter waxes (to keep the crusts pliable) that can ruin your entire home-baked dish. What a shame! To the nervous tart maker, never fear! Zan's Honeyed Tart Crust fool-proof, darn tasty recipe is here.

MAKES 1 9-INCH CRUST

1 large egg yolk
1 tablespoon heavy cream
1 tablespoon honey
½ teaspoon vanilla extract
1½ cups flour
⅓ cup confectioners' sugar
¼ teaspoon sea salt
8 tablespoons unsalted butter, cut into ½-inch pieces, chilled

Whisk the egg yolk, cream, honey, and vanilla together and set aside.

In a food processor, mix the flour, sugar, and sea salt until well combined. While the blades are running, add chilled butter, a few pieces at a time, in short pulses until the mixture resembles coarse cornmeal. Add the egg

and cream mixture through the feed tube and continue to process until the dough just comes together, about 10 seconds. Do not overwork the flour. Turn the dough onto a sheet of plastic wrap and press into a disk. Cover tightly with plastic wrap and refrigerate at least 1 hour to relax the dough.

Let the chilled dough soften at room temperature for about 10 minutes before rolling it out and fitting it into a 9-inch tart pan with removable sides. Set the dough-lined tart pan on a large plate and freeze for 30 minutes.

Preheat the oven to 375°F and adjust an oven rack to the middle position. Set the dough-lined tart pan on a baking sheet and prick the dough all over with a fork. Bake for 20 to 30 minutes or until the tart crust is a light golden brown. Transfer the tart pan to a wire rack to cool.

Bronx Bees' Rhubarb Tart

All I have to say is . . . orgasmic. If you're watching your weight, look at it this way. Eggs have protein and cream contains calcium. This may not technically be heart-healthy, but it is soul healthy! The Happiness Factor will go skyward when a cool spring drags on, and on, and on.

SERVES 6

1½ pounds crisp, tender rhubarb stalks
⅔ cup sugar, divided
2 teaspoons anise-flavored liqueur such as Pernod, ouzo, anisette, Sambuca, or Ricard
2 teaspoons finely grated orange zest
3 eggs
1 large egg yolk
¾ cup heavy cream
1 teaspoon sea salt
1 precooked Zan's Honeyed Tart Crust (see p. 36)

Wash the rhubarb and trim the ends. If the rhubarb is older and harvested later in the season, pull off the strings. Cut each stalk into 1 x ½-inch pieces and, in a glass or ceramic bowl, toss with ⅓ cup of the sugar. Cover and refrigerate for at least 3 hours.

Preheat the oven to 400°F.

Drain the rhubarb in a colander set over a large skillet. Press firmly on the rhubarb to extract as much juice as possible. Boil the juice over high heat until syrupy. Remove the skillet from the heat and add the rhubarb pieces, tossing to coat the rhubarb in the syrup. Add the anise-flavored liqueur and orange zest. Let cool slightly to room temperature.

In a medium bowl, whisk the eggs with the additional egg yolk, cream, sea salt, and remaining sugar.

Distribute the rhubarb evenly in the tart shell and pour half of the custard over the rhubarb. Let the custard settle down into the tart shell for a few minutes and then add the rest.

Bake the tart at 400°F until the custard is golden and bubbling, about 30 minutes. Transfer to a wire rack and allow to cool to room temperature. Serve at room temperature directly from the unmolded tart pan. Do not refrigerate the leftovers. I doubt you'll have any anyway.

21
Metro-North Railroad

La Finca del Sur

E. 138th Street & Grand Concourse, Bronx, NY 10454
(646) 725-2162 | www.bronxfarmers.blogspot.com

Annie Moss, cofounder and board member of La Finca del Sur, sits surrounded by green on a hand-hewn picnic table over the rumbling subway vents. The Metro North Railroad roars behind her shoulders, and the Grand Concourse Highway screeches behind mine as we enjoy the comfort of three acres of an urban farm oasis. Surprisingly large, La Finca del Sur urban farm is surrounded on all four sides and below by people traversing city arteries.

Entering its fourth growing season, this gigantic lot was once strewn with mounds of road waste and garbage, all buried beneath a chest-high thicket of overgrown weeds. Clearing the space with help from the Greenthumb NYC program, the women who built La Finca del Sur uncovered treasures of immense and ancient oaks and mulberry trees.

La Finca del Sur is a women-run urban farm in the South Bronx. This cooperative nonprofit is passionately dedicated to providing community leadership opportunities to Latina and black women and their allies. Choosing not to adopt the standardized bylaws commonly used by most social service providers, they instead chose an organizational structure that celebrates the value of diversity that promotes equality. Specific role-based job duties were created, which are filled by different women who all have equal value, equal influence, and equal organizational control. Additionally, new members are encouraged to take leadership positions, nurtured along and counseled by the multiethnic women all working together to create this revolutionary urban farm.

Officially La Finca del Sur is a cooperative committed to building healthy neighborhoods through economic empowerment, increased nutritional awareness, training, and education, and advocating for social and political equality and food justice in their low-income community.

Unofficially, this urban farm is a breath of fresh air and respite in this South Bronx

Annie Moss

neighborhood. On foot and having taken a wrong turn, I did in fact get lost for a while. I suffered through about two or three miles of extra walking in the blazing heat of a humid August day, lugging my fifty-pound backpack before I finally caught up with Annie Moss. I grumbled a bit at my unplanned misadventure, but in retrospect I believe it was the most important part of the journey to La Finca del Sur.

This neighborhood is the epicenter of commutes to all parts of Manhattan, Queens, the Bronx, and the northern counties, and the roads, ramps, bridges, and subways crisscross this area like a jumbled pile of pick-up sticks. There is hardly any room for anything green, and I became one of many peoples—with or without homes—squatting on haunches under an overpass for a brief reprieve from the sun, despite the amplified noise and blasts of diesel fumes.

Entering La Finca del Sur and sitting in the shade surrounded by the smells of sweet basil and hibiscus, I realized the extraordinary value that urban farming space provides. These multinational and multicultural and multilingual women have jointly created an Eden whose benefits sooth an aching humanity beyond just feeding its belly.

La Finca del Sur Caramelized Brussels Sprouts

Most people HATE brussels sprouts. I usually don't mind because that leaves more for me. With that said, most dislike brussels sprouts because they contain sulfur. Buying young, small sprouts that are freshly picked is key. Also cutting brussels sprouts in half enables that sour sulfur taste to escape during cooking. Without the sulfur, brussels sprouts are sweet, crunchy, caramelized morsels of goodness.

SERVES 4-6

¼ cup unsalted butter
1 pound brussels sprouts, stems removed, and quartered
2 tablespoons packed dark brown sugar
¼ cup lemon juice
Sea salt and freshly ground pepper
1 tablespoon olive oil
2 thin slices prosciutto, slivered

Melt butter in a sauté pan over medium-low heat. Add sprouts and sauté for 1 minute or until coated with butter. Turn heat to low and cook another 5 minutes, tossing occasionally, or until soft. Sprinkle in sugar and lemon juice. Turn the heat to medium-low and continue to cook, stirring occasionally, for another 8 to 10 minutes until sprouts are nicely caramelized. Season to taste with sea salt and pepper.

While sprouts are cooking, heat olive oil in a separate small skillet over medium-high heat. Sprinkle in prosciutto and sauté for 1 to 2 minutes or until crisp. Drain prosciutto on paper towels and set aside.

When sprouts are ready, sprinkle prosciutto over the sprouts and serve warm.

CAN IT!

Annie's Baked Cheese Hors d'Oeuvres with Cilantro Salsa

Someone told me that hors d'oeuvres are dead . . . that we're too busy to cook and serve food before cooking and serving food. Well, for those of you who still yearn for the *amuse bouche,* this is a great little recipe. If you're like me, I make the hors d'oeuvres and then have a glass of wine and then relax and forget to cook the rest of the meal. You can be like me. It's ok.

MAKES 12

Salsa

1 clove garlic, minced
¼ teaspoon sea salt
¼ teaspoon pepper
2 tablespoons olive oil
2 tablespoons whiskey or sherry vinegar
¼ cup sweet white wine
¼ cup finely chopped scallions
1 cup packed fresh cilantro leaves
1 or 2 jalapeños peppers, seeded and finely chopped
Sea salt and pepper to taste

Cheese Cups

¼ cup unsalted butter, softened
1 teaspoon Dijon mustard
2 cups grated cheddar, at room temperature
¾ cup flour
¼ teaspoon sea salt
¼ teaspoon cayenne pepper
2 tablespoons vodka
8 ounces pancetta or bacon, cooked, drained, and coarsely crumbled
Sea salt and pepper to taste

To make the salsa, mash garlic, sea salt, and pepper with a mortar and pestle until a thick paste forms. Transfer to a bowl and whisk in olive oil and whiskey until emulsified into a vinaigrette.

In a blender, combine wine, scallions, cilantro, and jalapeño pepper and pulse until well combined. Combine with garlic-whiskey vinaigrette and allow to stand at room temperature for at least 1 hour before serving.

Preheat oven to 325°F.

Coat mini muffin cups with nonstick cooking spray. Using an electric mixer, briefly beat butter with mustard, and then add the cheese. Beat until well combined and only small bits of cheese are still visible. Combine flour with sea salt and cayenne and stir into cheese mixture. Add vodka, 1 tablespoon at a time, until a dough forms.

Gather the dough into a log about 12 inches long. Cut in half lengthwise, and then cut each piece in half again. Cut each quarter into three pieces, ending up with 24 equal pieces of dough. Using your hands, shape each piece of dough into a cup-like shape and fit the dough into muffin cups, pressing to evenly fill each up right to the top.

Bake in the center of the oven for 20 minutes or until lightly browned. Cool on a rack for 10 minutes. Gently extract the cheese muffin cups. If cheese cups stick, turn pan up-side down over the rack and rap the bottom of each muffin cup with the back of a spoon until released.

When ready to serve, fill each warmed cheddar cup with cilantro salsa and top with pancetta or bacon pieces. Season to taste with sea salt and pepper. Serve immediately.

Rikers Island GreenHouse

148 W. 37th St., New York, NY 10018
(212) 757-0915 | www.hsny.org

Rikers Island is the largest jail complex in the United States, encompassing ten different jails within 415 acres of razor-wired space. On any given day, more than twenty thousand people call Rikers Island their home—some of whom are sentenced people who serve short terms and detainees who may or may not be sentenced to different durations. This warehouse for New York City's criminal justice system has a particularly uncommon ally: the Horticultural Society of New York.

The Horticultural Society of New York (HSNY) administers a jail-to-street horticulture program for men and women inmates. The two-acre urban farm at Rikers Island has greenhouses and attached classrooms through which the temporarily segregated community members can learn about plant science, ecology, horticulture skills, gardening construction, and design. They also build bird- and bat houses and bird feeders

and grow plants for gardens in NYC schools, parks, and neighborhoods. This exemplary programming provides valuable job training, works to re-create direct physical connections between people and nature in the city's urban communities, and fosters secondhand associations between those living in urban communities and their extradited neighbors.

In his book *Doing Time in the Garden,* James Jiler points out that "there is little evidence that a punitive jail environment reduces recidivism rates. What it does, rather, is instill a sense of futility which will accompany them as inmates return to their homes and communities." Prison horticulture both dramatically alters the jail's physical landscape and allows inmates an opportunity to directly interact with and exercise positive creative input within their surroundings. The act of planting food and flowers and trees becomes a personal experience, and the decisions of what to plant provide inmates an avenue of self-expression and empowerment. This humanizing of the Rikers Island facilities also calms the aggression and anger that are part and parcel of the prison system ethos.

Hilda Krus of the Horticultural Society is the acting director of the Rikers Island program. It started nearly thirty years ago as an idea between the president of the Horticultural Society and the commissioner of the Department of Corrections, who agreed that one of the causes

ALL PHOTOS THIS STORY BY HORTICULTURAL SOCIETY OF NEW YORK

of the high recidivism rate is that former inmates don't often have any alternatives to the lifestyle that brought them to Rikers Island in the first place. Often they lack skills to acquire jobs, and often they don't have hope. Former inmates find themselves thrown back into a society where they don't know what to do or where to go.

Incidentally, prison horticulture is not new. Initiated in the early nineteenth century, the penitentiary system as we know it today was a lasting gift of the Pennsylvania Quakers who believed a loss of liberty was better punishment than public torture, executions, and beatings. As part of the original penitentiary model, inmates were required to work in assigned details that would make the jail as sustainable as possible, doing such jobs as construction, maintenance and sanitation, tailoring, quarrying stone, food preparation, and farming.

Hilda specializes in horticultural therapy, which is a focused and goal-oriented activity that uses plants and plant materials to work with people through various conduits such as vocational training, cognitive learning, and enhanced physical learning, and it can also be used for mental and spiritual health rehabilitation. Hilda uses the example of horticultural therapy with stroke victims. "You might have a situation where one hand is weaker than the other, and you can use plants, such as working with seeds and transplants, to improve muscle coordination."

One focus for the HSNY GreenHouse program at Rikers is vocational training, providing skills to allow inmates opportunities for gainful

employment when they return to their communities. "People right NOW are on Rikers," Hilda says "but they're not going to be there forever. They'll come back. Some people come out sooner than others, but eventually everybody will return to society." Landscaping provides good jobs that don't exclude those who have been formally incarcerated. These vocational skills are hands-on, highly educational, with mandated class-time learning reinforced with quizzes and tests.

Hilda Krus

Happily, our current criminal justice system is far removed from the Quaker's "loss of liberty" principles, which advocated harsh, unsanitary, and inhumane conditions and long hours of forced manual labor with mandated work quotas. And now that we've addressed and fixed many of those problems, programming like the HSNY GreenHouse project is reinstating agriculture into the prison system this time as a vehicle of hope.

Teaching Rikers's residents how to learn is the first part of the journey together. "We do not underestimate our people," Hilda says. "In the beginning we often have inmates who are absolutely frozen before taking a quiz, absolutely petrified that they may do something wrong. Over time, we show our students that they are able to learn something—positive reinforcement that many haven't experienced in a long time. Somebody thinks they can learn! Most of our inmates start classroom learning not trusting themselves; over time they enjoy and appreciate being mentally challenged and succeeding."

Many, when first arriving at Rikers Island, are depressed, exhausted, and traumatized; they have gone through horrible situations, experienced great violence, or hurt themselves and

their families very badly with substance abuse. Hilda and the GreenHouse Program allow new inmates time and space to wind down and use the beauty of the gardens to take a deep breath and start coping again.

We can all draw some conclusions that plants and green things are, shall we say, Zen. But plants as therapy—therapy for people who, in some cases, have done insidious and inhumane things? If someone is sick, we send them flowers. If someone goes to Rikers Island, the HSNY and former inmates have created for them a garden. Why is this strategy so fundamentally important? Hilda says, "Inmates at Rikers are still members of our society. If we look at how we want them to come back out before rejoining our neighborhoods, we want them to maybe feel good enough about themselves and potentially be less angry or less depressed. We want to provide positive feedback through their having successfully learned and mastered new skills. We want them to know that they could have valuable jobs such as turning blight into beauty, necessary jobs that provide positive loveliness and succulent food to the NYC metropolis."

Hilda's Butter Beans with Sausage

Raise your hands if you would rather die than eat lima beans. Most would, but that's because most of us have only been exposed to dried lima beans and we're at a loss with what to do when a lima bean is fresh and recently picked. Butter beans are lima beans, and when fresh, are so tender they practically melt in your mouth—like butter. You gotta try this! If you don't, that's ok. There's more left for me.

SERVES 4

3 pounds fresh butter beans (lima beans) in pods
½ cup sliced dry-aged andouille sausage
2 tablespoons olive oil
1 cup seeded and diced red bell pepper
2 cups stock of your choice (see p. ix)
½ teaspoon sea salt
½ teaspoon pepper
½ teaspoon cayenne or hot pepper flakes
2 tablespoons lemon juice
1 tablespoon maple syrup
Sea salt and pepper to taste
Crusty bread for serving

Shell the butter beans and rinse well with cold water. Set aside.

In a large saucepan, sauté the sausage in olive oil over medium-high heat until browned. Add the red pepper and continue to cook until the pepper has browned and released all of its juices, about 5 to 7 minutes.

Stir in the stock, sea salt, pepper, cayenne pepper, and butter beans. Reduce the heat to low, and simmer, covered, for 45 to 50 minutes, or until the beans are tender. In a separate bowl, whisk together lemon juice and maple syrup. Stir into the bean mixture. Season to taste with sea salt and pepper and serve warm with crusty bread.

Rikers Island Fried Green Okra

You'd be surprised to know how many farmers are growing okra in the city. Okra is an edible member of the cotton (mallow) family and it isn't surprising that most, therefore, liken its taste and slimy seed-like texture to chewing on a dirty pair of shorts. Okra is an acquired taste. Deep frying okra is a nice baby-step introduction to this weird vegetable. You'll be amazed at how versatile and tasty okra can be.

SERVES 4

2 pounds fresh young okra
1 tablespoon sea salt
2 teaspoons pepper
⅛ teaspoon cayenne pepper
½ cup cornmeal
½ cup peanut oil for frying, divided
Sea salt and pepper to taste
Slices of fresh tomato for serving

Rinse the okra, drain well, and dry completely with paper towels. Cut off both ends of the okra and discard. Cut the pods crosswise into ½-inch pieces and sprinkle with sea salt, pepper, and cayenne pepper.

Pour the cornmeal into a large plastic bag. Add the okra and shake well to cover with cornmeal. Transfer to a large strainer or colander and shake gently to remove all of the excess cornmeal.

Pour half the oil into a large and heavy skillet and heat until hot. Add one-third of the okra and fry quickly, turning over with a slotted spoon, for 5 minutes or until golden brown. Remove the okra from the pan and drain on paper towels. Place on an ovenproof platter and keep warm in a 250°F oven.

Continue frying the okra in batches, adding more oil and allowing the oil to heat up again between batches. Season to taste with sea salt and pepper. Serve piping hot with slices of garden-fresh tomatoes.

Cabbage
'All Seasons Mix'

Wanaqua Family Farm

464 E. 136th St., Bronx, NY 10454
(212) 788-7928 | www.grownyc.org

Every school in New York City is going to get its own garden. One of these, the Wanaqua Family Garden, sits on a ten-thousand-square-foot lot next to the Jonas Bronck Elementary School. Started ten years ago by a neighborhood resident, Louis Rosario, this is an urban farm that is part of the Grow to Learn Program of GrowNYC, which facilitates and promotes the creation of a sustainable school garden in each and every public school across New York City.

Research shows, and it's also plain to see, that children today are increasingly disconnected from the world of natural and healthy food—a problem with hugely negative implications to their development and to the health and future of their environment. For all the bling and bright lights with which NYC sparkles, its schoolchildren struggle to connect with Mother Nature on a regular basis. These smart and savvy kids know their way through the complicated underground subway system but are often ignorant about where their food comes from. Backyards are rare. Expansive backyard gardens with rows of exotic vegetables are nearly nonexistent.

The NYC public education system includes 1,600 schools responsible for the feeding and education of more than 1.1 million children. School yards are often fenced in and paved with asphalt, and a surprising number of the city's young people are suffering from double-digit obesity and diabetes rates due to poor nutrition and lack of physical activity.

Gerard Lordahl is the current voice behind the Wanaqua Family Garden, one of GrowNYC's current bank of three hundred functioning school gardens, that has the dual purpose of serving two nearby schools. Additionally, this urban farm serves as a community garden, offering twenty available vegetable beds to the surrounding neighborhood.

Gerard Lordahl and GrowNYC Members

The Wanaqua Family Garden rivals any Westchester mansion's wraparound deck and manicured estate. Here community members gather for dinner, or stop in to read their books or the daily paper with coffee. At any given time, you'll find old men or families resting for a picnic, chatting on the phone, or taking a siesta.

During the school months, extensive food curriculum is provided for the school children, complete with chefs and recipes and large bowls of pounded pesto that the kids have made for their own lunch. Each school grows what would delight or interest its children the best.

To see corn growing in the Bronx is an absolute delight. For all of the negative press the corn plant has received from its variations of sweetener to oil, seeing it wave tall in the wind here is truly inspirational. I cheered.

Bronx Pizza Dough

There is absolutely NOTHING better than fresh and homemade pizza dough. I added a little zip to typically boring recipes by adding a bit of beer. If you're not necessarily a fan of pizza, then use this recipe to make quick rolls or pigs in a blanket or whatever your heart desires.

MAKES 1 10-INCH PIZZA CRUST

⅓ cup plus 2 tablespoons Pale Ale beer
¾ cup flour
½ teaspoon instant yeast
½ teaspoon dark brown sugar
½ teaspoon sea salt
4 teaspoons olive oil

Place the Pale Ale beer in a small saucepan and boil on high for 30 seconds to remove its alcohol and carbonation. Cool completely to room temperature.

In a medium-size bowl, whisk together the flour, instant yeast, and brown sugar. Make sure that there are no lumps. Whisk in the sea salt. Make a well in the center of the flour and pour in the Pale Ale. Using a rubber spatula or wooden spoon, gradually stir the flour into the beer until all of the flour is moistened and a dough just begins to form, about 30 seconds. The dough should be rough looking and not silky smooth. Do not overmix.

Drizzle olive oil over the dough and turn it gently to coat all sides of the dough and bowl with the oil. Cover tightly with plastic wrap.

If you want to prepare the pizza immediately, allow the dough to rest at room temperature for 1 hour or until doubled in size. For the best flavor, make the dough at least 6 hours and up to 24 hours ahead of time, and allow it to sit at room temperature for only 30 minutes until slightly puffy, then put the dough in the refrigerator and remove it 1 hour before you want to put it in the oven.

Preheat the oven to 475°F at least 30 minutes before baking and place the oven shelf at the lowest level.

With oiled fingers, lift the dough out of the bowl. Drizzle the oil remaining in the bowl onto the cookie sheet or pizza stone that the dough will be cooked on. Set the dough on the cookie sheet and press it down with your fingers to deflate it gently. Shape it into a smooth round by tucking under the edges. If there are any holes, knead it very lightly until smooth. Allow to rest for 15 minutes.

Using your fingers, press the dough from the center to the outer edge to stretch the dough into a 10-inch circle, leaving the outer ½ inch thicker than the rest to form a lip. Brush the surface of the dough with any remaining olive oil. Cover and allow to rest for an additional 30 to 40 minutes until it becomes light and slightly puffy with air.

Bake the pizza for 5 minutes at 475°F on the lowest rack. Remove from the oven, add your favorite toppings, and bake again until the crust is golden and any cheeses have melted, about 5 to 10 additional minutes.

Wanaqua Garden Rainbow Pepper Pizza

Pizza has been around forever. The term pizza first appeared in a Latin text in AD 997 claiming that a property tenant was to give the bishop of Gaeta twelve pizzas every Christmas day and another twelve every Easter Sunday. (I wish my landlord liked this idea.) The innovation that gave us what we call "pizza" today came from the eighteenth century when it was common for the poor people living in Naples to add tomato to their yeast-based flat breads.

MAKES 1 10-INCH PIZZA

2 tablespoons olive oil, divided
1 medium onion, thinly sliced
1 red bell pepper, cored, seeded, and cut lengthwise into strips
1 orange bell pepper, cored, seeded, and cut lengthwise into strips
Sea salt and pepper to taste
¾ cup crumbled sweet Gorgonzola dolce (or other blue cheese) or fresh mozzarella
2 tablespoons minced fresh basil leaves
Anchovy fillets (optional)

In a large sauté pan, heat 1 tablespoon olive oil over medium heat. Sauté the onion, stirring often, until the onions are fully browned. Remove to a separate bowl. In the same sauté pan, add the remaining tablespoon of olive oil and sauté the peppers, stirring often, until they are tender when pierced with a fork, about 15 minutes.

After the first 5 minutes of baking the Bronx Pizza Dough crust at 475°F (see left), remove the pizza from the oven and spread the onion evenly over the surface, leaving a narrow border uncovered all around the edge. Arrange the peppers in a spiral on top of the onions. Strew little bits of the Gorgonzola or other cheese on top and sprinkle with the fresh basil.

Return the pizza to the oven for 5 to 7 minutes or until the cheese is melted and the crust is golden. Using a spatula, lift a corner of the pizza to check to see that the bottom has browned.

Serve warm. Garnish, if desired, with anchovy fillets.

Chapter 3

MANHATTAN (NEW YORK COUNTY)

5-Boro Green Roof Garden

20 Randall's Island, Randall's Island, NY 10035
(212) 410-8910 | www.nycparks.org

The rooftop of the Five Borough (5-Boro) Technical Services Complex on Randall's Island is New York City's largest green roof. The Parks Department is testing more than twenty different growing systems side by side and recently added a four-thousand-square-foot vegetable garden. In the spring of 2007, the Five Borough Technical Services Division of the New York City Department of Parks and Recreation began a program installing green roofs of various designs on top of their own headquarters. This is the only known green roof in the world that features such distinctively different, side-by-side growing systems for research, learning, and comparison. Food is grown in elevated planters, overhead trellises, tray systems, pre-vegetated

Rick Gordon and NYC Parks Team Member

FDR Dr
2 Ave-125 St
Harlem Riv Dr
20

modular systems, tower hydroponic gardens, cedar planter boxes, and more.

A toothy-smiled and happy-go-lucky Rick Gordon is the 5-Boro Green Roof Garden's farmer maestro. It is agreed that his new toys, the organic Tower Garden hydroponic systems, are incredibly cool. This patented vertical gardening technology comes with stackable tower columns that utilize a closed-loop hydroponic circulation system that can grow between twenty and forty-eight plants on less than three square feet of earth. Rick oversees towers of tomatoes, lettuce, herbs, and strawberries. "The soil part of the farm [also growing tomatoes, lettuce, herbs, and strawberries for comparison] entails a lot more work in the beginning," Rick said. "You plant them, weed them occasionally, and watch them grow. The tower gardens require constant work to balance the pH. Each tower drinks more than five gallons of pH-balanced nutrient solution every day, but they're producing more food as a result." One of Rick's tomato towers had over seventy-five ripening tomatoes on it. I counted.

This building also boasts a sophisticated green-roof monitoring system and storm-water retention system incorporated into this mega outdoor classroom and working laboratory. A one-inch rainstorm will deposit approximately 10,600 gallons of water on the 5-Boro Green Roof gardens. Half of that is retained and absorbed by the various plants and vegetables, and the remaining 5,000 gallons are collected in water tanks and used to irrigate the garden during periods of low precipitation. The 5-Boro Green Roof Garden prevents more than 425,000 gallons of storm water from gushing down into the sewer system and out into the Hudson River every year. "This is just one roof," Rick said.

"Think of all the runoff pollution we could prevent if everybody had one of these."

Rick zips around the 5-Boro Green Roof Garden with all of the zeal of a kid rooting around in a toy chest. "Rooftop urban farms are so beneficial. They help cut down the carbon footprint from food transportation and also help reduce food transportation costs while absorbing vehicle emissions. If we had more urban farms in NYC, we could produce more food, clean the environment, and reduce the amount of traffic on the roads. What a no-brainer!"

This amazing garden stands next to the toll booths on the FDR Drive as it proceeds into the Triborough Bridge. Unless you know to look out of your windows as you pass through the E-ZPass lane, you might not see this glowingly chartreuse, food-growing green roof above the vehicle repair garage for the Parks Department. This rooftop garden is open to the public to visit, to students and universities for research projects, and to other farmers and urban food-growing venturers to come to study, learn, and select a system that will work best for their particular area.

"Urban farming is taking off like crazy! I can't believe the number of new farmers who are coming out here to study and learn from us and then going back to their neighborhoods and creating their own urban farms. I have a job at the Parks Department. Yet, every day I have the additional pleasure of inspiring hundreds of new farming projects throughout the city. Wow!"

Rick's Curried Cauliflower Soup

This basic soup recipe can be modified in a myriad of different ways. Primarily, it is fantastic served with a large dollop of fresh, plain yogurt. Also, feel free to add leftovers from other meals to make the soup a bit chunkier—such as roasted chicken or roasted root vegetables. This soup is also particularly good served like one would serve French Onion Soup—in a crock topped with dried bread and cheese and heated under the broiler. This is one recipe to play with.

SERVES 6

7 cups diced cauliflower florets
¼ cup olive oil, divided
Sea salt and pepper to taste
4 shallots, sliced
1 tablespoon minced ginger
1 tablespoon Joe Holzka Green Curry Paste (see p. 198), plus extra for serving (optional), or 2 teaspoons curry powder
4 cups stock of your choice (see p. ix)
2 teaspoons lemon juice

Preheat oven to 450°F. Toss cauliflower florets with all but 3 tablespoons of olive oil. Season liberally with sea salt and pepper and roast on a baking sheet, stirring occasionally, for about 20 minutes or until tender and browned.

In a large stockpot, sauté shallots, ginger, and curry paste in 3 tablespoons olive oil over medium heat. Add roasted cauliflower and toss to coat with the spices. Pour in the stock and bring to a boil. Reduce the heat to low, cover, and simmer until the cauliflower is very soft, about 10 minutes. Add lemon juice.

Puree the soup using a blender or immersion blender. Season to taste with sea salt and pepper. Serve in warmed soup bowls with an extra dollop of Joe Holzka Green Curry Paste, if desired.

5-Boro Roasted Sesame Asparagus

Asparagus is a sign of spring and the end of a cold, miserable winter! This dish is fabulous served with a bit of seared salmon or roasted portobello mushrooms on a sandwich (or cooled as a lovely accompaniment to a vodka martini).

SERVES 6

⅓ cup soy sauce
3 tablespoons rice vinegar
½ tablespoon plus 2 teaspoons toasted sesame oil
1 tablespoon honey
2 pounds fresh asparagus, tough, woody stems removed
2 teaspoons olive oil
½ teaspoon sea salt
Sea salt and pepper to taste

In a small bowl whisk soy sauce, rice vinegar, ½ tablespoon sesame oil, and honey together to make a vinaigrette.

Preheat oven to 425°F. Spread asparagus stalks in a single layer onto a cookie sheet and brush the asparagus with 2 teaspoons of sesame oil and olive oil and sprinkle with sea salt.

Roast the asparagus for 10 to 12 minutes or until tender and browned. Serve warm, drizzled with vinaigrette. Season to taste with sea salt and pepper.

BERKSHIRE BERRIES

1352 Main St., Becket, MA 01223
(800) 523-7797 | www.berkshireberries.com

David Graves lives with his family in Becket, Massachusetts, nestled deep in the Berkshires. Daily, these operators of Berkshire Berries are out of bed before 2 a.m. and make the three-hour trek through morning rush-hour traffic into New York City to start their day of farming on the tops of random, nondescript buildings where his bees live.

For years, David worked for his dad in the Berkshires. Now in his later years, he is finding his place and identity by growing food and making maple syrup and jams and jellies. "It's an independent feeling that is hard to describe. Making something yourself and being able to sell it is extremely rewarding—especially when people like it! It takes a lot of time to build up

David Graves

RHUBAR
Maple Syrup Tastes Better
HORSERADISH
JELLY
HOT GARLIC JELLY
Berkshire Berries
Berry Hot Garlic

your confidence when you're doing something like this," David says.

Specializing in top-bar beehives, which run horizontally, David is constantly experimenting with ways to make his bees as comfortable as possible by creating horizontal rather than vertical comb chambers. David's bees will make their way through the city in two miles in any given direction, forced to adhere to the man-made traffic patterns. They get a free pass on the traffic lights, however, as they zoom in, around, between, and alongside the infinite maze of brick and mortar buildings. In the back of a cab, if you looked up, and looked closely, you'd see nonhumans also commuting to work right above you.

Having watched *Animal Planet* and seen killing swarms of bees that terrorize neighborhoods and hunt small children, and having forgotten to refill my EpiPen, it was slightly unnerving to stand on a fifth floor rooftop next to large beehives in ninety-eight-degree weather. I realized that, like most, I am slightly afraid of these yellow things that fly.

Yet, without gloves, masks, or plumes of acrid smoke (which feigns an approaching brush fire and freaks the bees, thus setting them to gorge on honey to save the hive), David simply removes the cover of the hive and goes to work inspecting each rack of combs for signs of stress. I don't feel so bad about my bee phobia as David explains his apprehensions and trepidations throughout his early years of beekeeping. Having learned when to handle bees, when to respect their privacy, and how to commune with them symbiotically, it is anthropomorphically apparent that David has built stacks and boxes of cross-species friendships.

Throughout our interview, the bees didn't do what I expected, which was to viciously attack me as I ran screaming down the stairs. Instead, they simply built their castles and then flew out into traffic in search of more building blocks. Most of us would not use the words *docile* or *amicable* or *serene* to describe open, swarming hives of stinging insects. Yet that is what beekeeping with David on a rooftop is like. In his calming presence, I found myself eagerly straddling open hives with my camera thrust deep into the bowels of Apoidean catacombs, my nose inches away from the queen. This was unbelievably cool!

Perhaps it is the way he laughs, but David Graves reminds me a bit of a kindred Elmer Fudd. Unlike Elmer, however, this proprietor of Berkshire Berries has a pathological fear of bears. It's easy to see why he finds great solace on the scorching flats of New York City rooftops. There is a Zen-like atmosphere way up here—the screeching grumblings of impatient drivers below muffled by the contented humming of these legions of garden pollinators in David's hives.

Berkshire Berries' Pomegranate Sangria

Warning to the wise and unwise alike: This is good. Darn good. It'll also knock your socks off. WARNING: Do not drink this sangria if you do NOT intend to speak your mind to your meddling mother-in-law at dinner.

MAKES 8–10 DRINKS

1 cup vodka
¼ cup honey, plus additional 2½–3½ tablespoons
1 750-ml bottle fruity, semisweet red wine
2 cups pomegranate juice
Juice of 1 orange
¼ cup pomegranate seeds
1 cup red grapes
Ice cubes

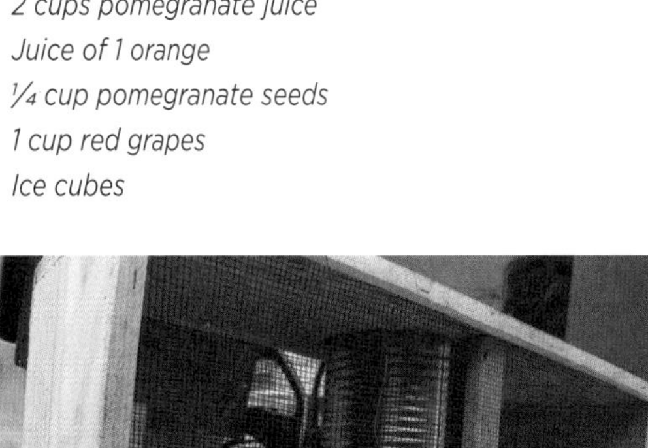

Mix vodka and ¼ cup honey in a large glass pitcher. Stir to dissolve the honey. Add remaining ingredients and stir to combine.

Place in refrigerator for at least 2 hours before serving. To serve, drizzle ice cubes with remaining honey warmed and top with Pomegranate Sangria.

David's Crumpets with Clotted Cream and Honey

I often wish I could go to a restaurant and order crumpets (the edible kind). If you're an old fan or a newbie to these English delights, I hope you enjoy this recipe! To amuse you while you're cooking, in Europe the term "crumpet" is used to describe a woman of certain below-standard moral character the same way we use the word "tart" as slang over on this side of the ocean. Enjoy the play on words.

SERVES 6

1 cup flour
1 teaspoon instant yeast
½ teaspoon sugar
½ teaspoon sea salt
¾ cup scalded milk plus 2 tablespoons, cooled to room temperature, divided
2 tablespoons melted unsalted butter, divided
¼ teaspoon baking soda
Clotted cream and honey

Find six 4-inch baking rings at least ½ inch high, or you can use six large tuna cans with tops and bottoms removed, thoroughly washed to remove any odors.

In a large mixer bowl beat together the flour, yeast, sugar, and sea salt. Use a paddle attachment if using a standard mixer. Add ¾ cup scalded milk and gradually increase the mixer speed. Beat for about 5 minutes until completely smooth and elastic. If not using a mixer, use a whisk to completely incorporate the flour, yeast, sugar, and sea salt until completely smooth and elastic, about 10 minutes.

Scrape the batter into a bowl that has been oiled with 1 tablespoon melted butter. Cover and allow the batter to rise until doubled, about 1 hour.

In a small bowl, mix together the baking soda and remaining 2 tablespoons scalded milk. Stir into the batter (this will deflate it again). Cover and allow the batter to rise until almost doubled again, about 30 minutes. The batter will be filled with bubbles.

Heat a heavy skillet over low heat until a drop of water sizzles when added.

Brush the skillet and inside of the rings lightly with the remaining melted butter. Set the rings in the skillet. Use a ladle or large spoon to pour the batter into the rings, filling them about two-thirds full. The batter will rise to the top during cooking. Cook over low to medium heat for 10 minutes or until they are nicely browned underneath. Turn the crumpets (and rings) with a spatula and continue cooking until browned on the second side.

Remove the crumpets from the skillet and from their rings and cool on a wire rack. Serve warm immediately or retoasted later with clotted cream (substitute crème fraîche or unsweetened whipped cream, if unable to find clotted cream) and copious drizzles of golden honey.

Dickson's Farmstand Meats

Chelsea Market, 75 9th Ave., New York, NY 10011
(212) 242-2630 | www.dicksonsfarmstand.com

I have to say that Jake Dickson is a good sport. I stumbled upon Dickson's Farmstand Meats by chance and instantly assaulted him with a drive-by, drop-in interview request. He looked suspiciously at the girl with a fifty-pound backpack, sweaty shoes, and knees smeared with dirt from crawling around for photographs and said, "Okay." I'm really glad he did.

More than just a butcher shop, Dickson's Farmstand Meats is reminiscent of a time when going to the local market was a normal thread in the social fabric of our neighborhoods and where "What's fresh today?" was the common greeting from a customer. Jake reaches deeply into who we are and makes us ache longingly for local butcher shops and a time before billions of hamburgers were sold from brightly lit, antiseptic drive-through windows, before the concept of consuming meat meant buying something wrapped in plastic on a Styrofoam tray from a big-box grocery store where not a blade of grass can be seen for miles, and before the word *cow* was replaced by the phrase "daily protein requirement." Thanks to Jake Dickson, city slickers now have access to meats from local New York farmers who are raising animals that are as respected and treated as gently during slaughter as they were out in the pastures.

Jake Dickson

Jake Dickson is not a butcher. He's a conscientious capitalist who operates one of the most successful and principled butcher shops in all of New York City. An honorable business, guided by ethical, moral, and pro-environmental guidelines, that is financially successful is a rare anomaly. All of Dickson's Farmstead Meats come from approved local farms that Jake visits himself. His rules are strict and clear: no antibiotics except to treat illness, no feedlots, no hormones, and no animal by-products in the feed. "I want to be able to say with complete comfort," Jake explains, "that all of the animals that I sell are not just high quality but have been treated extremely well throughout their lives and treated humanely at the end."

SHORT RIBS
9/lb

GRASS-FED
CHICKEN STEAK
14 /lb

Whole animals are brought to Jake and his team of butchers. All of the butchering, carcass processing, charcuterie, smoking, curing, and Master stock making happens in the store, and they do more whole-animal butchering than any other company in New York City. As opposed to the secrecy surrounding commercial meat processing, everything is butchered on full-customer display at Dickson's Farmstead Meats. "My industry has strayed very far from what you would consider healthy and good practices," Jake said. "Feedlots, overuse of antibiotics, and growth hormones are rampant. Small local farmers who ethically produce animals for human consumption are fundamentally important. They provide traceability and accountability. If the general American public could see what happens in the wider meat industry because of scale and profit motive, 99 percent of them would no longer have any interest in eating that food."

As a chef, I can say that Jake's bacon has taken the number two spot on my list of Best Bacon Ever. (The Dickson's Farmstand Meat chorizo claims the coveted number one spot on my list of perfect charcuterie, and an artisan from another country that I haven't yet visited will have to compete fiercely to best Dickson's creamy smoked *lardo*.) If you do buy anything, take home a container of Jake's velvety, gelatinous stock. This is the real deal, and you'll start to wonder why we're so easily snowed into buying thin, watery containers of pretend-a-stock in aseptic boxes for exorbitant prices from commercial food companies.

The American environmentalist Aldo Leopold once said, "All ethics rest upon a single premise: that the individual is a member of a community of interdependent parts. The land ethic simply enlarges the boundaries of the community to include soils, waters, plants, and animals." Dickson's Farmstand Meats brings one into the rare presence of skilled butchers who excel at the tangible fabrication of ethically raised local meats. "Our farmers are everything," Jake said. "Everything."

Dickson's Chorizo Stuffing

Turkey stuffing with sausage and cauliflower? You might think I'm crazy but this is absolutely a perfect combination with any roasting fowl—be it duck, goose, chicken, turkey, or any of the multitude of edible birds in-between. Too, it's nice to have a vegetable-laden stuffing instead of just, and only, bread. Enjoy!

MAKES 9 CUPS

8 cups dried ciabatta bread cubes
1 red pepper, seeded and diced
2½ teaspoons grapefruit zest
½ cup minced fresh Italian parsley leaves
2 teaspoons sea salt
¾ cup olive oil
2 leeks, white part only, sliced thin
2 cups diced cauliflower florets
1 grapefruit, peeled, sectioned, and diced
1 pound dry-aged chorizo sausage, sliced thin
3 teaspoons fresh thyme leaves
1-1¼ cups stock of your choice (see p. ix)
Sea salt and pepper to taste

Place bread cubes, red pepper, grapefruit zest, parsley, and sea salt in a large bowl. Stir to combine.

Heat olive oil in a large, wide skillet over medium heat. Sauté leeks, cauliflower, and grapefruit cubes until grapefruit releases its juices and the cauliflower begins to brown. Add sliced chorizo and continue to cook until it is mostly cooked through and crumbling, about 3 minutes. Add thyme leaves and sauté for about 30 seconds more. Add mixture to the bread cubes and stir to combine.

Add stock, ½ cup at a time, to moisten the bread cubes. The stuffing should be moist but not soggy. Season to taste with sea salt and pepper; err on the side of slightly over seasoning the mixture.

Just before roasting the turkey, place the stuffing into the cavity and back of turkey without packing it too tightly. Or place the stuffing in a 6-quart greased, covered baking dish and bake at 325°F until hot, about 45 minutes.

Jake's Lamb Loins with Chocolate Sauce

For some crazy reason lamb and chocolate go perfectly well together. What's nice about this recipe is that both while you're cooking (and also after the meal), you can continue to enjoy the health benefits of chocolate and Madeira (which is what I usually do.) Although the recipe calls for ½ ounce of chocolate, you better buy a lot more than that.

SERVES 4

1 tablespoon black peppercorns
2 boneless lamb loins, about ½ pound each
2 tablespoons sea salt
2 tablespoons olive oil
1 tablespoon finely chopped garlic
1 cup stock of your choice (see p. ix)
½ ounce dark bittersweet chocolate, 70% (or more) cacao

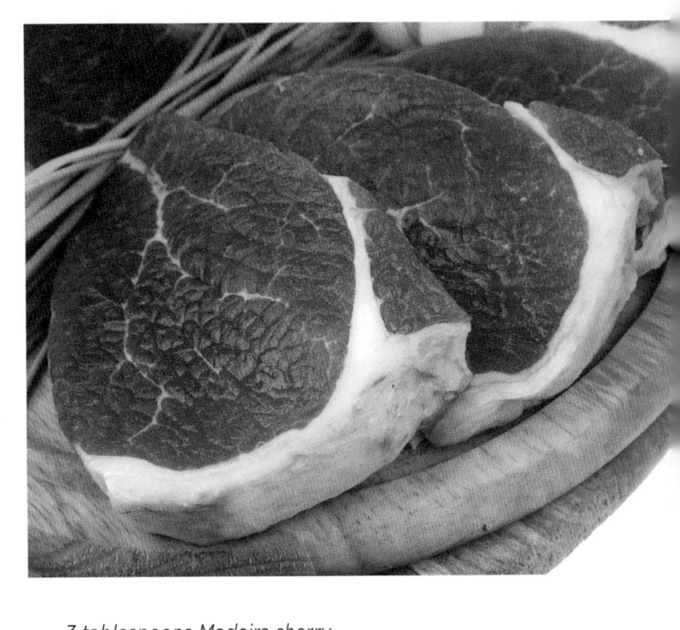

3 tablespoons Madeira sherry
Sea salt and pepper to taste

Preheat oven to 250°F.

Coarsely crush the peppercorns with a mortar and pestle. Sprinkle the pepper over the lamb loins and season with sea salt.

Heat the olive oil in a large frying pan until very hot. Add the lamb and cook for 5 to 6 minutes per side until evenly browned. Transfer to a baking dish and keep warm in the oven.

Remove the pan from the heat, add the garlic, and stir. Return the pan to low heat and cook, stirring until the garlic just beings to color. Pour in the stock, stirring to deglaze the pan. Bring to a boil and cook until the stock is reduced by half. Add the chocolate, stirring vigorously until melted. Add the sherry and season to taste with sea salt and pepper.

Slice the lamb into thick slices and serve with the sauce.

Fifth Street Farm Project

600 E. 6th St., New York, NY 10009
(212) 477-1735 | www.theearthschool.org

Abbe Futterman is the cofounder of the Earth School, a progressive elementary school whose mission reflects Manhattan's deep commitment to the sustainable uses of the earth's resources. She is a science teacher turned agricultural visionary turned farming instructor through her science curriculum at the Robert Simon Complex, a massive public school building that houses elementary schools P.S. 64 and the Earth School, and the Tompkins Square Middle School. Her Fifth Street Farm Project addresses childhood obesity, storm-water runoff, and climate change. Conceived by a grassroots organization of teachers, parents, and green-roof advocates, the Fifth Street children's education farm will expand from its fenced quarter-acre container garden (accessible only by crawling through a window) to a three-thousand-square-foot urban farm roof deck with spectacular views of Manhattan.

The Fifth Street Farm Project is led by the National September 11 Memorial & Museum designer Michael Arad. In other states surrounding New York, getting even one small garden plot into a school is a five-year advocacy project demanding public hearings, raucous PTA councilwomen, and near riots. Their success rates are minimal, and the process is frustrating. Yet, a three-school collaborative, sky-high organic vegetable garden in the East Village is something that Abbe Futterman calls "normal."

Each and every one of Mrs. Futterman's students plays in the garden during science class. As the rooftop garden is completed, all students at the Robert Simon Complex will enjoy a fresh-food-growing experience, along with outdoor dirty garden recreation time, making fruit and vegetable experimental discoveries, and taking cooking lessons using their own hand-grown produce. This experience will continue for ten continuous years from prekindergarten through to the end of eighth grade. Abbe is the science

Abbe Futterman

teacher for all grades and sees each student continuously throughout their learning years. "I look at the curriculum longitudinally," Abbe says. "What are they doing in terms of botany, entomology, and anatomy? I try to find some plant-based and garden-based lessons for that portion of the curriculum." Abbe knows children occasionally learn more by playing. "They started digging through the compost heap and found seeds such as peach pits or melon seeds." In response, Abbe created a Free Planting Zone where each student can plant the random seeds they've found and watch them grow. "I also put out a big box of 'mystery seeds,'" Abbe said. "I want them to have free rein to play, explore, and see what comes up."

And in fun and exciting ways, all things farming and food are integrated as value-added learning opportunities throughout the school's math, English, social studies, and all other class syllabi. Forget about the boring question of Charles traveling from St. Louis at eighty miles an hour and Chuck traveling from Florida at forty-five miles an hour and where these two shall meet. The National Curriculum Standards for History Studies might still include random facts about the Dust Bowl, but under the Earth School's tutelage, students might actually learn what a fallow field is and how crop rotation and cover crops could have prevented such devastating wind erosion.

These students are likely to develop greater empathy for the Okies who were forced to become migrant farm workers, and, in turn, have a better cultural and historical understanding of literary works based on the events of this historical natural disaster like *The Grapes of Wrath* and *Of Mice and Men*. Thus in contrast to schools that send kids to history class to learn random facts and then to English class to read random literature, the Fifth Street Farm Project is using food as an underlying framework to unite the random facts of a standardized curriculum.

The practice of implementing gardens at schools for purposes of hands-on education has been used in parts of Europe since the early 1800s. Nearly one hundred years later when school gardens were built in America, they were included in the educational curriculum with the added agenda of correcting the perceived social and moral ills at the height of the Temperance Movement which advocated for women and children's rights and social protections. Most of the burgeoning school garden networks were spearheaded by women in urban settings on the East Coast as part of the Progressive Movement which

sought civic and social alternatives to industrialization. Fast-forward almost another one hundred years to 1997 when Alice Water's Edible Schoolyard program was born in Berkeley.

Hailed as a school garden pioneer, reaction to Alice Waters is split. She occasionally draws a great deal of both praise and criticism. At the heart of the Edible Schoolyard movement is a garden-based curriculum, paid for by Alice's private foundation that addresses a wide spectrum of poverty's ills and, as Alice says, "helps students learn the pleasure of physical work." This split reaction to Alice Waters is probably due in large part to the fact that this Waters-led resurgence of school-based agriculture is based on personal responsibility and meaningful work. This is in sharp contrast to our historical school garden programming which was founded on self-sustainability, women's rights, children's rights, and community cohesiveness and collaboration.

The Fifth Street Farm Project is also a school garden, but its purpose and rationale are extraordinarily different. To start, this urban farm is a collaborative effort both requested and collectively built by teachers, parents, community members, and industry professionals from all over Manhattan. Second, the Fifth Street Farm Project is a "value-added" environmental experience not based on a puritanical values agenda. Also, the demographics of these East Village school children are not comparable to those in Berkeley, California. These students are average kids, growing up in a typical New York neighborhood, going to a normal school, but learning extraordinary things beyond the basics.

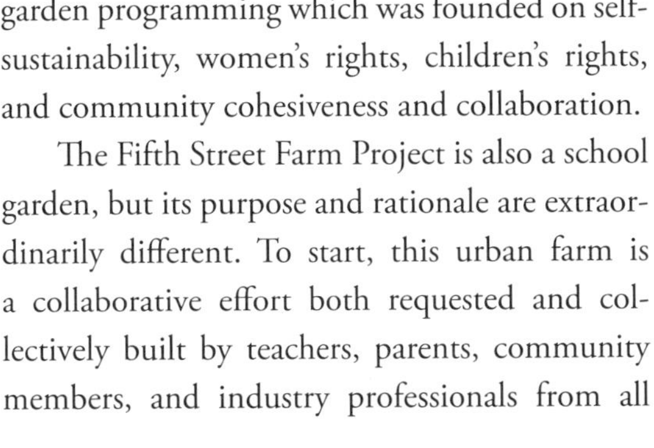

Instead of being forced to stay inside all day sitting on hard wooden chairs under fluorescent lights, children welcome the chance to crawl out of Abbe's window for fresh air as a welcome change of pace. "Our whole school is based on active and project-based learning," affirms Abbe. "All my kids are always so disappointed when we can't get outside into the garden. They look forward to this all day."

Abbe's Apricot Swiss Chard

If you're making this recipe in school, use apple cider or apple juice instead of whiskey. Anyway, all of the alcohol gets burned off in the cooking and what remains is a nice flavor that helps marry the apricots to the chard. This is great on toast, as filler for lasagna or other stuffed pasta recipes, or cold in pita bread with hummus. YUM!

SERVES 4-6

1 cup sliced, dried apricots
3 tablespoons whiskey, other alcohol, or water
1 pound swiss chard
3 tablespoons unsalted butter
1 cup thinly sliced leeks
½ teaspoon ground cumin
½ teaspoon sea salt
½ teaspoon pepper
3 tablespoons heavy cream
Sea salt and pepper to taste
Lemon wedges

In a bowl, soak the sliced apricot pieces in whiskey. Set aside.

Rinse the swiss chard thoroughly in cold water and dry in a salad spinner. Trim the ends by about 1 inch, cut the swiss chard into 2-inch pieces, and set aside.

Melt butter in a large skillet over medium heat and sauté the leeks, stirring until golden, about 2 minutes. Add the swiss chard, apricots and whiskey, cumin, sea salt, and pepper, and continue to cook until the swiss chard is soft and the excess liquid has evaporated. Add cream, toss to combine, and cook until the cream just starts to boil. Season to taste with sea salt and pepper. Serve warm with lemon wedges.

Fifth Street Roasted Tomatillo Salsa

Kid friendly, teacher approved. To make the salsa less hot, use a green banana pepper or a red bell pepper.

MAKES 2 CUPS

1 garlic clove, peeled
4 medium tomatillos, husked, rinsed, and halved crosswise
1 jalapeño chile, seeded, stemmed, and roughly chopped
2 tablespoons orange zest
⅓ cup loosely packed diced cilantro leaves
½ cup water
2 shallots, minced
Sea salt and pepper to taste

Place garlic and tomatillos, cut side down, in a 10-inch skillet over medium high heat. When the tomatillos are well browned, after about 5 minutes, flip and brown the other sides of the tomatillos and garlic.

Scrape the garlic and tomatillos into a blender and let cool to room temperature. Add the jalapeño, orange zest, cilantro, water, and shallots. Puree until smooth. Season to taste with sea salt and pepper. Pour into a jar, secure the lid, and refrigerate until ready to use. Shake vigorously before using.

Randall's Island Children's Learning Garden

20 Randall's Island, Randall's Island, NY 10035
(212) 830-7722 | www.grownyc.org

Currently, only a handful of New York City public schools have an active school-garden program. That's where the Randall's Island Learning Garden comes in. A six-thousand-square-foot urban farm in the heart of the Randall's Island Sports Foundation Complex provides schoolchildren the opportunity to grow, harvest, and eat garden-fresh produce. A majority of children arrive by bus from the Bronx, East Harlem, or other underserved communities. These kids, who otherwise don't have access to open spaces, are provided with the opportunity to race freely across lush open fields to their heart's content. Phyllis Odyssey, director of horticulture at the Randall's Island Park, together with her assistant Eunyoung Sebazco, provides a mind-blowing garden learning experience to New York City's underserved youth.

"Carrots are our most popular vegetable," Phyllis notes. "Almost every kid has eaten a carrot. But when we pull them out of the ground, they're flabbergasted! Some even squeal and run away. These realizations can be, as they say, 'freaky' in the beginning. It's amazing how disconnected we are from our food supply. Surprisingly enough, once they get used to the idea that food grows in the dirt, the kids dig right in and chow down on everything we harvest together. We don't just slap cleaned vegetables on sterile plates and tell them to eat. We involve them in the entire process of growing, harvesting, and preparing food. If you've made something, you kind of want to taste it. Getting kids to eat vegetables is surprisingly easy."

The Randall's Island Children's Learning Garden is made up of forty-two raised beds and twenty in-ground gardening plots. It grows lacinato kale, leeks, purple and green broccoli, cheddar cauliflower, lettuce, beans, peppers, cucumbers, squash, heirloom tomatoes,

Eunyoung Sebazco and Phyllis Odyssey

The Rice Seeds are then separated from its panicle by beating them, or Threshing
More Threshing

eggplant, artichokes, carrots, beets, beans, corn, and rice. Yes, rice. There is a fully functional rice paddy adorned with green and white lily pads. A special sensory learning area of the garden provides food and plant experiences beyond mere food cultivation. There is a Garden of the Four Senses where children can feel prickly rosemary, rub soft and velvety sage, and crush lemongrass for perfume while smelling the different scents of these plants on their skin. Most public spaces have a "look, but don't touch" social etiquette. On Randall's Island, everything has been built to foster the sensory experiences. There are saltwater marshes and freshwater marsh habitats that can be explored on elevated boardwalk bridges, fruit trees to climb, and a million different flowers to touch and smell.

Technically part of the borough of Manhattan, Randall's Island is a mega sports complex with Ichan Stadium, sporting fields, and a golf and tennis center. Amid all these things, the island offers a variety of planted and themed gardens. Bicycle and pedestrian pathways run along the island's entire waterfront past the wildflower meadows. The Wetlands Stewardship Program and the Oyster Garden Program serve the fish, birds, and wildlife of the nine acres of wetlands. Amid all this, Phyllis and Eunyoung happily tend the Children's Learning Garden and the minds of the children who visit. "Here, unlike many other public parks, Randall's Island is still a blank canvas," Phyllis says. "This is a wide-open dirt palette for us to create anything we desire. Of all that we do, we particularly love the children's garden. As a horticulturist, I can grow anything and make anything look beautiful. Growing a garden to specifically educate kids about food and nutrition is extremely rewarding in very different ways."

Randall's Island Pounded Pesto

This is a GREAT outdoor activity for kids and adults alike. If you're working with kids, add a few drum sets, gourd rattles, and other noise makers preferably in the shade. Pound away and make sweet pesto music. If you're making this with adults, serve cocktails first. The garlic is optional, and the recipe is geared for school environments cued to be extra cautious of nut allergies.

SERVES 4

1 garlic clove, sliced thin
2 cups packed fresh basil leaves
¼ cup fresh parsley leaves
1 tablespoon lemon zest
7 tablespoons olive oil
½ teaspoon sea salt
⅔ cup grated Parmesan cheese
Sea salt and pepper to taste

Add the garlic to a small, dry skillet and toast over medium heat, stirring frequently, until golden and fragrant, about 2 minutes.

Combine basil and parsley in a heavy-duty, gallon-size ziplock plastic bag. Pound the bag with the flat side of a meat pounder or rolling pin until all the leaves are bruised.

Add the garlic, pounded herbs, lemon zest, oil, and sea salt to the bowl of a food processor and process until smooth, stopping to scrape down the sides as necessary. Transfer the mixture to a small bowl, stir in the Parmesan, and adjust with sea salt and pepper.

Serve as a condiment or as a glaze for grilled meats or vegetables. To use with pasta, thin with ¼ to ½ cup of reserved pasta cooking liquid.

Phyllis & Eunyoung's Fresh Spun Pasta

Pasta often gets a bad rap. True, flour is high in carbohydrates. With that said, up to 25 percent of the 2 cups of flour—so ½ cup—can be exchanged for healthier ingredients such as whole wheat flour, nut or bean flours, such as garbanzo bean flour, or other nutrient-rich grain flours. Making pasta is as simple as pie and a fun way to have guests participate in making the meal. You can also add up to 1 tablespoon of additional flavor, such as lemon juice or spice paste or, if you want to add dry spices, add 1 teaspoon of a spice of your choice. Curry is good! If you're in a funky mood, add food coloring!

SERVES 6

2 cups flour
¼ teaspoon ground white pepper
3 large eggs, beaten
1 tablespoon sea salt

Pulse the flour and pepper in a food processor fitted with a steel blade to aerate it evenly. Add the eggs and process until the dough forms a rough ball, about 30 seconds. If the dough is too crumbly, add water ½ teaspoon at a time, or if too wet, add flour 1 tablespoon at a time, and process until the dough forms a ball.

Turn the dough ball onto a dry work surface and knead until the dough is smooth, about 1 to 2 minutes. Cover with plastic wrap and allow to rest for at least 15 minutes or up to 2 hours.

Using a manual pasta machine, roll out the dough as directed.

Bring 4 quarts water to a rolling boil in a large pot. Add sea salt and pasta to the boiling water and stir to separate the noodles. Cook until al dente, about 5 minutes. Reserve ½ cup of the pasta cooking liquid. Drain the pasta and return it to the pot.

If using Randall's Island Pounded Pesto (see p. 77), stir ¼ cup reserved pasta-cooking water into the pesto in a separate bowl. Whisk to combine. Toss the hot pasta with the thinned pesto. Add additional pasta cooking water as needed.

Riverpark "Pop-up" Farm

430 E. 29th St., New York, NY 10016
(212) 925-6900 | www.riverparkfarm.com

A pop-up farm is growing amid the skyscrapers and bustling streets of Manhattan on a stalled construction site of the future West Tower of the Alexandria Center for Life Science. The term *pop-up* comes from the newly popular trend of pop-up, temporary restaurants. These roving supper clubs can be created in any space for any length of time without permanency. Riverpark Farm is the first-ever pop-up farm. Built with soil-filled milk crates and tangles of water lines, this urban farm is completely moveable and transportable, allowing it to shift around the Alexandria Center as subsequent phases of construction begin.

Chef Sisha Ortúzar

The Alexandria Center for Life Science—New York City is New York State's first life-science park and complex designed to foster unique and innovative collaborations among New York's world-renowned research scientists, academics, and medical institutions. Manhattan is the site of Alexandria's worldwide flagship location. Two additional buildings are set to be completed by 2017, but, as with more than six hundred other stalled construction sites throughout New York City, they await a better financial climate. The Riverpark Farm is an ingenious pop-up urban farm that is temporarily repurposing just one of these open areas and creating local economic activities that benefit the environment and beautify the neighborhood. When construction begins again, the farm will simply be moved.

Wherever it is actually going to be at any given time, Riverpark Farm is a fifteen-thousand-square-foot black-box food-growing machine supplying Tom Colicchio's Riverpark Restaurant. The restaurant's chef and founding partner, Sisha Ortúzar, is clearly having a lot of fun playing with his food. Most chefs spend a majority

of their time indoors in cramped, hot spaces. In Manhattan, this also means spending a majority of time below street level. If they aren't sweating under a grease hood, then they're waiting for the delivery trucks to bring thousands of pounds of food in boxes and then spending hours in frigid coolers and freezers packing and storing and sorting and rotating this food. In contrast, Chef Sisha gets to run outside and pick okra.

It's amazing what one cubic foot of soil in a black milk crate can grow. The only things that can't be grown at Riverpark Farm are corn and field grains. Their 3,300 milk crates grow an impressive 243 tons of food in rotational plantings. "The different crops come in waves. Last week we picked over one hundred pounds of cucumbers each day. This week we're in the midst of eggplant craziness," Chef Sisha said. "Because of that, our menu changes accordingly. Our customers love it. They're very interested in knowing about seasonality and how we're going to specialize the menu options as these vegetables are ready to be harvested." In addition to preparing stocks and demi-glace and marinades indoors, Chef Sisha spends his time in the fresh air and sun cultivating his other basic culinary components, such as typical greens and herbs, pattypan squash, tomatoes, peppers, melons, strawberries, edible flowers, and more. This dual indoor/outdoor food prep by the chef makes for an unparalleled culinary experience.

Tom Colicchio, Sisha Ortúzar, and the Riverpark Farm team are revolutionizing the entire restaurant world. Instead of "farm to table,"

this restaurant offers a "crate to plate," locally grown, recently harvested, grown-on-premises, harvested-by-the-chef dining experience that serves handcrafted food in unexpected flavor combinations. If you think that's a mouthful, try eating there. Nothing is more enjoyable than savoring an exquisite feast prepared by a culinary artist.

Riverpark Farm Zucchini Salad with Ricotta on Grilled Bread

Riverpark Farm grows six varieties of basil (Thai, lime, cinnamon, globe, opal, and sweet) and both spearmint and peppermint, so they like to combine two or three varieties of basil and mint for this dish. The thickly sliced bread is especially delicious in this recipe if you can use a grill to char the bread.

Recipe provided by Chef Sisha Ortúzar and Riverpark Farm.

SERVES 4

4 medium-size zucchini or yellow squash
4 tablespoons olive oil, divided
1 small onion, julienned
1 teaspoon sea salt, divided
1 teaspoon freshly ground black pepper, divided
1 Fresno chile, thinly sliced
Juice and zest of 1 lemon
Salt and pepper to taste
4 thick slices rustic country or sourdough bread
1½ cups sheep milk ricotta
6 squash blossoms, petals separated
½ cup loosely packed whole mint leaves
½ cup loosely packed whole basil leaves (and flowers, if you have them)
½ cup shredded Parmesan cheese

Slice two of the zucchini about ¼-inch thick. Heat 2 tablespoons of olive oil in a pan over medium heat, add the zucchini slices and onion, and cook until lightly browned on both sides. Season with ½ teaspoon each of sea salt and pepper, then set aside to cool.

Steady a mandoline over a medium bowl and slice the remaining zucchini into thin ribbons. Add the chile and season with the remaining olive oil, lemon juice and zest, and sea salt and pepper to taste. Gently toss the salad, adjusting the seasoning if necessary, then let it sit for a few minutes.

Meanwhile, char or toast the bread and spread each slice generously with the fresh ricotta.

Add the squash blossoms, mint, and basil leaves to the salad and toss again.

Place the cooked zucchini evenly onto each toast, then top each with the salad. Garnish with Parmesan and basil flowers, if available. Season to taste with remaining salt and pepper.

Riverpark Farm Oyster Tacos

Recipe provided by Chef Sisha Ortúzar and Riverpark Farm.

MAKES 8 TACOS

Spicy Mayo

1 large egg yolk
½ teaspoon Dijon mustard
½ teaspoon chopped garlic
1 teaspoon Espelette pepper
1 tablespoon white wine vinegar
1 cup grapeseed oil
1 tablespoon water, plus additional if needed
¼ cup extra-virgin olive oil, divided
½ teaspoon kosher salt
½ teaspoon freshly ground black pepper

Tacos

1 large green, unripened tomato, sliced very thin
1 teaspoon kosher salt
1 small red onion, julienned
Juice of 2 limes
Salt to taste
8 large or 16 small fresh oysters
Canola oil for frying
1 cup cornmeal
8 fresh white corn tortillas
1 jalapeño chile, seeded and thinly sliced
½ cup minced fresh cilantro leaves
Pepper to taste
4 radishes, cut into wedges
1 lime, cut into 8 wedges

To make the mayo, place the egg yolk, mustard, garlic, Espelette pepper, and vinegar in a blender or food processor. Start the blender, slowly add half the grapeseed oil in a thin, even stream, and then add 1 tablespoon water. Slowly add the remaining grapeseed oil until incorporated, adding another 1 teaspoon of water if it seems too thick. Slowly add the olive oil until smooth and season with kosher salt and pepper.

To make the tacos, toss the green tomato with kosher salt in a bowl and let it sit for 30 minutes. Cover the

onion with the lime juice, season with salt, and let it sit for 30 minutes.

Meanwhile, shuck the oysters into a bowl along with their juices.

Heat about 2 inches of canola oil in a small pot to 350°F.

Dredge each oyster in the cornmeal, shake off any excess, and gently place them into the oil. Fry one or two oysters at a time for about 35 seconds, or until the outside is crispy. Drain on a paper towel and season with salt.

Drain the tomatoes. Warm the tortillas, then spread a generous amount of Spicy Mayo onto each. Add an oyster (or two), then top with some tomato, onion, jalapeño, and cilantro. Season to taste with salt and pepper. Serve with a few radish and lime wedges on the side.

URBAN FARM AT THE BATTERY

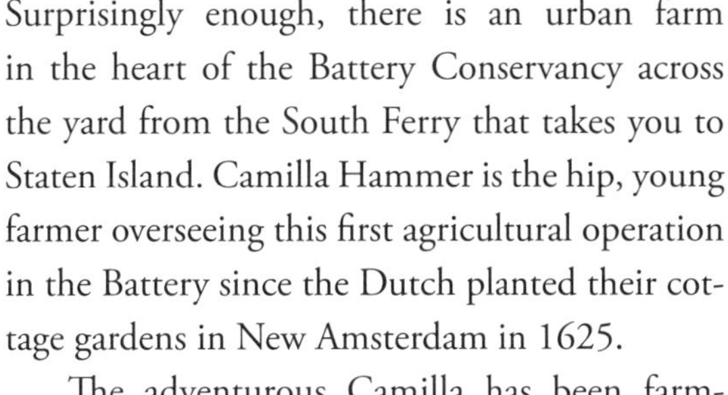

1 NEW YORK PLAZA, NEW YORK, NY 10004
(212) 344-3491 | WWW.THEBATTERY.ORG

Surprisingly enough, there is an urban farm in the heart of the Battery Conservancy across the yard from the South Ferry that takes you to Staten Island. Camilla Hammer is the hip, young farmer overseeing this first agricultural operation in the Battery since the Dutch planted their cottage gardens in New Amsterdam in 1625.

The adventurous Camilla has been farming all over the world, but this is her first official posting in an urbanized setting. "This is a dream come true," she said. Her mission at the Urban Farm is to provide an outdoor classroom, to promote healthy and conscientious eating, and to foster a sense of community in this neighborhood of glass skyscrapers, towering corporate giants, and international business influencers. A farm-market stand sits in the shade beneath these mega buildings for those who do not wish to rent plots to grow their own food, and more than 680 kindergarten through twelfth-grade students from eight different schools come to learn, grow, and get their hands dirty in the soil. Schools are growing a wide variety of sumptuous foods, such as eggplants, carrots, husk cherries, pumpkins, and summer savory. Many students return to their cafeterias with vegetables in hand to incorporate this student-grown produce into the making of their school lunches.

The playful turkey-shaped Urban Farm at the Battery was recently named one of NYC's top-five urban farms. Its design was inspired by Battery Park's long-time resident turkey, Zelda, and it's shaped and protected by five thousand repurposed bamboo poles in the shape of Zelda's silhouette. The posts were donated by artists Mike and Doug Starn following their rooftop exhibition *Big Bambu: You Can't, You Don't, and You Won't Stop* at the Metropolitan Museum of Art, and the urban farm's protective fencing was designed by local artist Scott Dougan. It is fitting that identical-twin artists Doug and Mike Starn's walkable fifty-foot bamboo artwork would find a repurposed home as a soil-protective fence in the heart of the financial district. Bamboo is incredibly strong and, like the produce grown at the Urban Farm at the Battery, bamboo constantly changes depending on the light and weather.

PS 3 The John Melser Charrette School
2nd & 3rd Grade - Charley
URBAN FARM AT THE BATTERY

The students who tend plots at the Urban Farm also keep us informed of their progress through their blogs. Sixteen-year-old Anika and her classmates reported planting carrots, tomatoes, sugar snap peas, hollyhocks, melons, and marigolds. Student Farmer Emma said, "I cannot believe how quickly this class went by. I really miss it! I am so happy that everything is beautiful and blooming now!" Student Farmer Rebecca reflected, "The experience at the farm has been most refreshing. I have enjoyed taking care of the plants and watching them grow, and farming in between my other classes has been a grounding and stress-relieving way for me to feel more connected."

This temporary urban farm that stretches across the Great Lawn of the Battery Conservancy was the brainchild of the savvy environmental club at Millennium High School. Its current site has been dug and redug several times to make ever-expanding underpasses for the Brooklyn Battery Tunnel, and the farm will exist through the end of 2012, when the area will once again undergo construction to become part of the Battery Conservancy's new bikeway and perennial garden. I suspect, however, this form of education-based urban farming will not disappear but will no doubt be shifted to a new location within the Battery Conservancy. "It's been too much of a great and positive thing for this neighborhood to walk away from," Camilla said.

Camilla's Uncooked Apricot Ice Cream

Ice cream can come in many different forms—why not try one with a nut-based milk! You'll love the flavor on a hot summer day! This is considered "raw" ice cream and is a nod to raw foodists!

MAKES 5 CUPS

½ cup dates, soaked in water to cover for 20 minutes
1 quart almond milk
7 fresh apricots, peeled and pitted, plus additional sliced fresh apricots for serving
1 teaspoon almond extract
1 peeled banana
1 tablespoon lemon juice

Drain the dates and discard the soaking water. Combine all ingredients in a blender and blend well. Chill the mixture for 1 hour and then place the mixture in an ice-cream maker and process until frozen. Serve with sliced fresh apricots.

The Battery Beet & Barley Risotto

Typically, risotto is a class of Italian dishes slowly cooked in broth over a long period of labor-intensive time. While this recipe is not "technically" a risotto because it doesn't require intensive stirring and long cooking times, it is a nice mock risotto that uses healthier grains. This can be a one-pot meal that is perfect served while wearing heavy sweat pants. It can also be a nice accompaniment to anything off the grill or out of a roasting pan. Its festive color makes for great visual feasting!

SERVES 6-8

1 cup pearl barley
3 cups stock of your choice (see p. ix)
1 cup water
1 tablespoon unsalted butter
1 medium red onion, chopped
1 clove garlic, minced
½ teaspoon ground allspice
1 teaspoon sea salt, divided
1 teaspoon ground pepper, divided
½ cup dry white wine
2 cups peeled, diced beets
¼ cup dried currants
2 tablespoons freshly squeezed orange juice
2 tablespoons freshly squeezed lemon juice
Sea salt and pepper to taste
¼ cup crumbled soft feta cheese

Rinse barley in a sieve under running water, stirring to make sure it is well rinsed, until water runs clear. Drain well and set aside.

Heat stock and water in a saucepan or in the microwave until steaming. Cover and keep hot.

Melt butter in a large saucepan over medium heat. Add the onion and sauté for about 3 minutes until starting to soften. Add garlic, allspice, and ¼ teaspoon each sea salt and pepper and sauté an additional 2 minutes until the onions begin to brown. Add the barley and sauté, stirring constantly, until the barely begins to turn brown and roast but not burn. Immediately add the wine and deglaze the pan, stirring constantly, until the wine is almost evaporated.

Stir in 2 cups of the hot stock mixture and the beets and currants. Cover, reduce heat to medium-low, and boil gently, stirring occasionally, for 25 minutes or until the beets are almost tender. Add a little more stock, if necessary, to keep the mixture moist.

Uncover and simmer for about 20 minutes, or until the barley is tender with a slight bite and the beets are tender, stirring often and adding stock, a ladleful at a time, as the previous addition is absorbed. Adjust the heat as necessary to keep the pot at a steady simmer. You may need to add a little more hot water if the mixture gets too thick before the barley is tender.

Stir in the orange and lemon juices. Season to taste with sea salt and pepper. Serve sprinkled with feta cheese.

FUDGE

Chapter 4

Orange County

Acorn Hill Farmstead Cheese

65 Red Barn Rd., Walker Valley, NY 12588
(845) 800-4239 | www.acornhillfarmsteadcheeses.com

Acorn Hill Farm has a fantastic six-duck welcoming and entertainment committee. "Ducks just struck my fancy," Joyce Henlon said. "They keep us amused as they really do sound like they're laughing sometimes." To test this theory, I made a rather defamatory remark about of one of our former presidents. "Wha wha wha," the ducks replied and bobbed their heads. This, of course, sent Joyce into giggles, and I felt like a world-class stand-up comedienne.

Acorn Hill Farmstead Cheeses is a small-scale family-operated microcreamery in the Catskill/mid-Hudson region of New York nestled on the border of Orange and Ulster Counties. Following organic practices and having taken the Northeast Organic Farming Association of New York farmer's pledge, Joyce Henlon produces some of the best goat cheese, goat fudge, and soap available—right in her six-acre backyard.

Joyce calls Acorn Hill Farmstead Cheese "a wonderful series of accidents." Namely, she found out that she loved goats. Secondly, a local farmers'-market master convinced her that having cheese at one of the market stalls each week would be a wonderful thing (hint hint). No slighter remark has ever led to such an epic journey. Building a micro-creamery is no small feat. First of all, milking, milk storage,

Joyce Henlon

and pasteurization equipment tends to be of a size and scope to accommodate large dairy herds that send their milk to be pasteurized in outside facilities. Second, handling milk from live animals entails serious risk of infection by deadly food-borne illnesses. It took a while to convince the state of New York that small backyard milking operations could be safe if the proper Occupational Safety and Health Administration (OSHA) standards are utilized and the food safety laws are followed. Dairy farms are among the most hazardous and deadly work environments in the United States. OSHA standards are rules and regulations that help keep farmers and farm workers safe on the job while creating and providing edible products that are safe for the consumer to eat.

In addition to the welcoming committee of ducks, the Acorn Hill Farm has eighteen goats that are all named and specially tagged with ribbons or bows or funny dog-collar charms. While Neo, Persephone, and Daisy all enjoyed exploring the camera, it was clear to see that Joyce has the same special attributes of most backyard livestock farmers. The smaller the agricultural scale, the more personal and intimate the relationships

are with the animals. Her farm produces cheeses, but its Joyce's eighteen "family pets" that help her do so. "I talk to the goats all of the time," she said. I decided to try it. "So," I asked, "are you pleased with the current state of the economy?" They all responded with "naaaaaaa" and shook their ears. A baited question, I know. Joyce answered, "I joke with people all of the time. The one thing about farming is that you may not always get a paycheck, but at least you'll always have a job. It's awfully hard to get fired by a goat."

The Acorn Hill Farm is a relaxed and comfortable space. The ducks swim in the baby wading pool and rest under the shade of the rainbow-colored patio umbrella. The goats are content to explore the back fields and eat weeds to their hearts' content. The newly born goat kids wrestle each other in straw in their safety play pens. The milking parlor is nestled comfortably in a small red barn. Extensive gardens produce herbs for the cheese, berries for jams and jellies, and vegetables for the family's dinner. "I wanted raw organic dairy products for my own family. After that, I've turned this into a business," Joyce said. "I am tremendously grateful for what I do. It's something that is purposeful, and I'm celebrated and supported by all of my neighbors. It's such a good feeling."

Joyce's Fennel Custard

Custards are perfectly acceptable at the dinner table. We tend to always serve custards as sugar-filled desserts, but think outside of the ramekin!

SERVES 4–8

2 shallots sliced
1 lime, peeled, sectioned, deveined, and minced
2 tablespoons unsalted butter
1 head fennel, trimmed, cored, and cut into quarters
1 tablespoon sea salt
3 large eggs
1 teaspoon fennel seeds
2 teaspoons grated lime zest
½ cup whipping cream
½ teaspoon sea salt
½ teaspoon ground white pepper

Preheat oven to 350°F.

In a small saucepan, sauté shallots and lime pieces in butter until the shallots start to caramelize and turn brown and the lime has released all of its juices. Set aside.

Place fennel in a large pot of boiling water salted with the sea salt and boil until crisp-tender, about 10 minutes. Drain and return the fennel to the hot pot and place over low heat for about 1 to 2 minutes to evaporate any remaining water. Cool to room temperature.

Place fennel in a blender with eggs, fennel seeds, lime zest, sautéed shallot mixture, and cream. Puree until smooth. Season with sea salt and pepper.

Divide mixture between four or eight ramekins, depending on how many you're serving. Place ramekins in a large roasting pan and fill the pan with boiling water to reach halfway up the ramekins' sides. Place the pan in the oven and bake for 30 to 35 minutes or until the custards are set and slightly puffy. Cool at room temperature for 5 minutes before serving in the ramekins or turning out the custard onto serving dishes.

Acorn Hill Grilled Portobello Sandwiches with Creole Salsa

Serve this with beer and classic rock-and-roll at your next BBQ cookout. If you're cooking hamburgers too, make extra salsa.

SERVES 6

Creole Salsa

- *2 ripe large tomatoes, peeled, seeded, and finely chopped*
- *2 green onions, minced*
- *¼ cup minced fresh parsley leaves*
- *¼ cup olive oil*
- *2 tablespoons white wine vinegar or white wine*
- *½ teaspoon sea salt*
- *½ teaspoon sweet paprika*
- *3 tablespoons minced fresh oregano leaves*
- *½ teaspoon freshly ground black pepper*
- *¼ teaspoon red pepper flakes to taste*
- *Sea salt and pepper to taste*

Mushrooms

- *¼ cup olive oil*
- *2 teaspoons freshly squeezed lemon juice*
- *6 portobello mushrooms, stems removed, caps wiped clean*
- *1 teaspoon sea salt*
- *1 teaspoon ground black pepper*
- *8 ounces fresh goat cheese, sliced into equal 6 pieces*
- *Sandwich bread and fresh greens (optional)*

To make the salsa, thoroughly combine all the ingredients together in a medium-size bowl. Cover tightly and allow to rest at room temperature for at least an hour.

In a small bowl, whisk olive oil and freshly squeezed lemon juice together. Brush over cleaned portobello mushrooms and sprinkle both sides of each liberally with sea salt and pepper.

Grill mushrooms gill-side down over medium heat, turning once, until tender, about 16 to 18 minutes. While they are still on the grill, distribute the fresh goat cheese to the upward facing gill side of the mushrooms. Cover the grill and allow the cheese to melt for about 1 minute. Remove from heat.

Drizzle each mushroom with Creole Salsa and serve immediately as an open-faced sandwich or with a handful of fresh greens between slices of your favorite bread.

Jones Farm

190 Angola Rd., Cornwall, NY 12518
(845) 534-4445 | www.jonesfarminc.com

Jones Farm is an agritourism destination of epic proportions. This isn't just any old standard orchard. Since 1914 the Jones Farm has supplied the Hudson Valley with homegrown produce, freshly baked goods, and the finest gifts and decorative home accessories. The country store is a full-service grocery store of locally grown products from the region's local farmers. They also carry an assortment of gourmet grocery items, including products from Robert Rothschild Farm, Stonewall Kitchen, Road Pasta, Green Mountain coffee, Dean's Beans Organic Coffee, Hoboken Eddie's Specialties, and D. L. Jardine's Foods. This is all in addition to the pièce de résistance: Grandma Phoebe's Bakery.

Where to start? First, it's best to pour a cup of hot coffee, because the moment you walk through the door you'll be assaulted with the

David, Catherine, Doris, and Belding

smell of freshly baked cookies, muffins, cakes, and pies. You'll moan along with everyone else when devouring the chocolate truffle cookies and be semicomatose after licking up every crumb of the legendary Deadly Carrot Cake. All that locally grown beta carotene covered in a semi-sweet frothy frosting with just a hint of sweetened cream cheese surely does the body good.

After you've stuffed yourself to your heart's delight and loaded the week's groceries into the car, you can peruse the Clearwater Art Gallery, featuring resident artist Terri Clearwater; take a fine art painting, drawing, or sketching course; shop in Catherine's fantastic gift shop; or go to visit Miss Fern the pig. The old Dutch gambrel dairy barn has been converted into a gift, decorative, and personal-accessory store that features children's toys, jewelry, candles, art prints, and other items drawn from more than five hundred local vendors.

Miss Fern, nicknamed "Dumpling," came to visit the Jones Farm from upstate New York and decided to stay. She is the most celebrated member of the Jones family, and folks come from miles around for her autograph and a photograph or two. Her best good buddy is David Clearwater, head proprietor of the Jones Family Farm. Three generations still work behind the counters of the general store.

While this is still a working farm producing fruits and vegetables, David and his father, Belding, have changed the farming operations to accommodate changes in agriculture and customer needs and demands over the years. Jones Farm, surrounded by the Schunemunk Mountain State Park on one side and the Black Rock Forest preserve on the other, was first a

Miss Fern

dairy farm, then a poultry farm, and now is an eighty-five-acre orchard and veggie farm with a tourist-destination twist. "When I was growing up," David says, "this area was a blue-collar working community, and we all had local jobs. Now, everyone lives here but works in New York City and commutes back and forth. We've had to grow from a neighborhood farm store to a full-service 'stop in after work on your way home' grocery store. My grandparents would be laughing out loud—'You're a farmer, for gosh sakes. What are you doing selling pancake mix?' But I have to say, thank god that we don't rely on just selling our vegetables and apples. This year, with Hurricane Irene, we wouldn't have been able to pay the bills. I feel so incredibly sorry for our neighbors. Everything was ready to come in for the harvest, and it just floated away. It was awful."

One thing that hasn't changed is the deep familial roots that keep this farm operational. David's mom, Doris, works in the bakery and country store, his wife, Terri, runs the Clearwater Art Gallery & Frame Shop, his daughter Catherine manages the Clearwater Gift Store, and David's dad, Belding, can still be found outside on the farm chipping wood, or feeding the animals, or otherwise puttering in helpful ways. It is a beautiful thing.

Miss Fern's Olive-Oil Carrot Cake

Carrot cake, or as otherwise known in my house, is vegetable cake! Add olive oil and dessert becomes a heart healthy meal. Frosting is optional, depending on how sweet you like your vegetable cakes. This also pairs perfectly with a very dry Riesling.

SERVES 10-12

Cake

2 tablespoons unsalted butter, melted
2½ cups flour
1¼ teaspoons baking powder
1 teaspoon baking soda
1¼ teaspoons ground cinnamon
¼ teaspoon ground nutmeg
⅛ teaspoon ground cloves
3 tablespoons grapefruit zest
½ teaspoon sea salt
3 cups peeled, shredded carrots
¾ cup packed dark brown sugar
¼ cup maple syrup
4 large eggs
1½ cups olive oil

Frosting

6 ounces cream cheese
¼ cup ricotta cheese, drained overnight
5 tablespoons unsalted butter, softened but still cool
½ teaspoon vanilla extract
2 teaspoons freshly squeezed grapefruit juice
1 cup confectioners' sugar

To make the cake, adjust the oven rack to the middle position and preheat the oven to 350°F. Brush the sides and bottom of a 9 x 13-inch baking pan with melted butter. Line the bottom of the pan with parchment paper cut to fit and brush with 1 tablespoon unsalted, melted butter as well.

In a medium bowl, whisk together the flour, baking powder, baking soda, cinnamon, nutmeg, cloves, grapefruit zest, and sea salt. Add the shredded carrots and toss to combine. Set aside.

In a food processor, pulse the brown sugar, maple syrup, and eggs until frothy and thoroughly combined. With the machine running, add the olive oil through the feed tube in a steady stream. Process until the mixture is light in color and well emulsified, about 30 seconds.

Scrape the mixture into a large bowl. Add the flour and carrot mixture and stir until completely mixed and no streaks of flour remain. Pour into the prepared baking pan and bake for about 35 to 40 minutes or until a toothpick inserted into the center of the cake comes out clean. Cool the cake on a wire rack for about 2 hours, or until the cake has cooled to room temperature.

To make the frosting, process cream cheese, ricotta cheese, butter, vanilla extract, and grapefruit juice until smooth. Add the confectioners' sugar and process until smooth and creamy, about 30 seconds.

Run a paring knife around the cake in the pan to loosen it. Invert the cake onto a serving platter and carefully remove the parchment paper. Spread the frosting evenly over the surface of the cake, cut into squares, and serve.

Jones Farm Tarte Tatin

During the height of apple season, we'll occasionally look around and say, "Ok, now what?" This is one recipe that works for breakfast, lunch, dinner, brunch, afternoon snack, or midnight munchies. For some happy reason, apples and caramel work perfectly served either alongside eggs or under ice cream.

SERVES 8

Zan's Honeyed Tart Crust (see p. 36), unbaked
8 tablespoons unsalted butter
⅓ cup dark brown sugar
1 lemon, peel zested into large, thin strips
1 teaspoon peeled and mashed fresh ginger, pushed through a garlic press
6 large Granny Smith apples, peeled, cored, and quartered
1 cup heavy cream
⅓ cup chilled sour cream

Make Zan's Honeyed Tart Crust pastry, but do not bake. Roll out into a 10-inch round, cover tightly with plastic wrap, and chill in the refrigerator on a large baking sheet for about an hour.

Preheat oven to 375°F.

Melt the butter in a 9-inch ovenproof skillet. Remove from the heat and sprinkle in the brown sugar, lemon zest, and ginger. Arrange the apples in the skillet in concentric rings. The apples will be very close together; keep the core side down in the pan as you shingle and stack all of the apples side by side in one layer.

Return the skillet to high heat and cook until the juices turn from butterscotch to a rich amber color, about 10 to 12 minutes. Remove the skillet from the heat and, using a fork, gently turn the apples over to caramelize the other side. Return the skillet to high heat and continue cooking and boiling to caramelize the second side and remove any trace of juices, about 5 minutes longer.

Remove the skillet from the heat. Slide Zan's Honeyed Tart Crust off the baking sheet and carefully lay it over the skillet. Without burning your fingers, tuck the dough edges gently up against the skillet wall covering the entire pan with tart dough.

Bake the tarte until the crust is golden brown, about 25 to 30 minutes. Set the skillet on a wire rack to cool for about 20 minutes before sliding a knife around the edges of the skillet to loosen the crust. Place a serving platter or large plate on top of the skillet and, holding the plate and skillet together, firmly invert the tart onto the serving plate. Scrape out any apples that remain in the skillet and put them back into place.

Beat the heavy cream and sour cream at medium-high speed with an electric mixture until the mixture thickens and holds soft peaks. To serve, accompany each wedge of apple tart with a generous dollop of the creamy whipped topping.

Lynnhaven Nubians

414 Church Rd., Pine Bush, NY 12566
(845) 744-6089 | www.lynnhavennubians.com

Lynn Fleming has been raising purebred Nubian goats since 1989. With extraordinary zeal, the Lynnhaven Farm has also jumped full steam ahead into raising poultry. A cacophony of cackles signals rescued peacocks and heritage turkeys racing each other up and down the driveway in search of the best bugs. Ducks and chickens of all sizes and colors are just about everywhere.

Lynn hand-feeds all of her baby goats and hand-milks all 130 of her mamas. You can find her creamy hand-whipped goat cheeses at nine NYC Greenmarkets. "There's obviously no rest for the wicked," Lynn jokes. "The truth is I do all of this work just to keep my goats. My primary goal is to keep the lights on and the goats fed. I think customers think it's a line, but it's not.

Lynn Fleming

These animals are truly amazing, and I love having them here with me." In her on-farm creamery, Lynn makes chèvre, feta, and ricotta goat cheeses and, occasionally, dry aged cheeses. She also sells goat meat. "It's not my favorite job," she said, "but it's part of being a farmer. I'm committed to my animals, and that also means that I'm committed to knowing what happens to them throughout the entire duration of their life cycles."

Why goats? Good question. Lynn worked with horses all of her life. Before Lynn started her own farm, she worked for another farmer. This former boss arrived home from a convention one day and announced that the farm would be getting goats because it was becoming agriculturally popular. Lynn said, "No. You are. It's your farm, not mine. Goats stink and I don't want anything to do with them." Nevertheless, Lynn and her boss started driving around to buy goats from other farms. "The rest is history," Lynn chuckled. "I decided I liked goats more than horses, and my boss hated them and decided not to buy goats after all. It was pretty funny," she mused. Lynn sold her last remaining horses and moved to Pine Bush, a small town on the border of Orange and Ulster Counties, with twenty goats and fifteen sheep to start her own agricultural operation. The sheep are gone, but she stuck with the goats. "Sheep are unbelievably stupid. It's almost embarrassing. If you have one or a hundred, they only have one

414

brain, and it's yours. Goats, on the other hand, think that they are smarter than you. Chances are that is occasionally true. If there's trouble, they jump right into the fray. I can't imagine life without them."

In a fantastically bright Hawaiian-style sleeveless shirt, Lynn navigates with contentment around her pens of animals large and small, feathered and hoofed. Her cherry-red fingernails clasp a logoed gas-station coffee mug, and she veritably sings with delight as we spread morning greetings to all things not human. It's clear that at Lynnhaven Nubians animals rule the roost. A disordered sense of higgledy-piggledy seems genuine and sweet, and in a kindly way. All of the farm equipment is older than Lynn herself but has been recycled, reconditioned, and set back to its proper use. Lynn's laid-back style and dedicated passion for animals is a lovely mix for a midlife lady farmer. It's obvious that this breast cancer survivor doesn't let too much slow her down or get her down.

Lynnhaven's Oatmeal with Cinnamon Figs

Really, what is more boring than oatmeal? Malt-O-Meal or Cream of Wheat, maybe. Oatmeal is not something that most chefs will include in their cookbooks simply because of the yawn factor. With that said, Lynn is a saucy gal. If anyone can spice up boring oatmeal, she can.

SERVES 3-4

3 cups plus 2 tablespoons water
1 cup heavy cream or milk
3 tablespoons unsalted butter
1 cup steel-cut oats, rinsed
1 cinnamon stick
¼ teaspoon sea salt
1 cup dried figs, stemmed and quartered
2 tablespoons maple syrup
¼ teaspoon vanilla extract
⅛ teaspoon ground cinnamon

Bring 3 cups water and the milk to a simmer in a large saucepan over medium heat. Meanwhile, in a large skillet, melt the butter over medium heat and sauté the oats, stirring constantly with a wooden spoon until golden and toasted, about 2 minutes. Stir the toasted oats into the simmering liquid, add the cinnamon stick, and reduce the heat to medium-low and simmer gently until the mixture thickens and resembles gravy, about 20 minutes.

Add the sea salt and continue simmering, stirring occasionally, until the oats absorb almost all of the liquid and the oatmeal is thick and creamy, about 10 minutes. Remove from heat and let the oatmeal stand, uncovered, for 5 minutes.

In a separate small saucepan, bring the figs, maple syrup, 2 tablespoons of water, vanilla, and ground cinnamon to a boil over medium-high heat. Cook until the liquid reduces to a glaze and the figs soften, about 5 minutes.

To serve, remove the cinnamon stick and heap mounds of steaming hot oatmeal into large bowls. Serve with cinnamon fig topping.

Farmer Fleming's Roasted Goat

Coffee and cumin is a match made in heaven. This recipe works wonders for any kind of roast and the smell of roasting coffee coming out of your oven will put you in a fantastic mood, indeed.

SERVES 4-6

2 large fennel bulbs, rinsed and cut into 8 wedges
1 large onion, sliced into rings
1 cup dried Turkish apricots
3 tablespoons olive oil, divided, plus additional for serving
3 teaspoons ground peppercorns
3 teaspoons sea salt
1 teaspoon cumin seeds
1 teaspoon ground coffee
2 cloves garlic, sliced
1 leg of goat, with or without the bone
3 tablespoons minced fresh parsley for garnish
Sea salt and pepper to taste

Lightly coat the fennel, onion, and apricots with 1 tablespoon olive oil and place at the bottom of a roasting pan that has a meat rack insert.

Using a mortar and pestle, mash peppercorns, sea salt, cumin seeds, coffee, and garlic into a rough paste. Add 1 tablespoon olive oil and continue working into a smoother paste. Add last tablespoon of olive oil to create a slurry.

Over a baking sheet, rub the coffee slurry into the leg of goat and allow to marinate, covered, at room temperature for about 1 hour.

Preheat the oven to 400°F. Place the goat on a meat rack, inside the roasting pan, and over the fennel, onions, and apricots.

With a bone-in leg of goat, bake at 400°F for 10 minutes, reduce heat to 375°F for 10 minutes, 350° for 15 minutes, and then 325° for 30 minutes until the meat registers 140°F on a thermometer.

With a boneless leg of goat, bake at 400°F for 10 minutes, reduce heat to 350°F for 10 minutes, and 325° for 30 minutes until the meat registers 130°F on a thermometer.

Remove from the oven, place the goat on a cutting board, tent with foil, and allow it to rest at room temperature for 20 minutes. Remove the meat rack from the roasting pan and increase the oven temperature to 425°F. Stir the vegetables at the bottom of the pan, and continue to roast for another 10 to 20 minutes until starting to turn dark brown and caramelize.

To serve, slice the goat and serve over caramelized vegetables. Drizzle with a dash of olive oil and top with a sprinkle of parsley. Season to taste.

Note: Bones stay cold and require lots more time in the oven to warm up, and the meat surrounding the bone needs more time to cook. This is the reason for the difference in cooking time and temperatures for the bone-in versus the boneless leg of goat.

Pierson's Evergreen Natural Beef Farm

1448 New York 211, Middletown, NY 10940
(845) 386-1882 | www.piersonsfarm.com

Way out in the middle of an extensive and fertile pasture lies a dilapidated old camper scrapped for junk. Or so it seems. Upon closer inspection, this is one helluva tricked-out mobile chicken coop. Joel Salatin was made famous by Michael Pollan in *The Omnivore's Dilemma* for his wood-and-mesh mobile chicken coops, and Orrin and Jackie Pierson do have his models of coops on their farm as well. But this Holy Rollin' Chick Mobile? This takes moveable chicken coops to Paris Hilton extremes! The fact that this refurbished camper was originally made by the Mallard RV Company is an additional dose of hysterically fowl humor to stick in one's craw.

Jackie and Orrin Pierson, proprietors of Pierson's Evergreen Natural Beef Farm, are two of the kindest and most down-to-earth people you'll have the great pleasure of meeting. When Orrin tells you that his cattle are NEVER crowded into feedlots or never receive unnatural, high-grain diets, he isn't lying. The Piersons' farm has retained the quintessential old-school look and feel of the eighteenth century. This farm looks exactly like those pictured on glossy "all natural" grocery store labels falsely promoting anything but a family farm. The rolling green fields stretch as far as the eye can see, while silos belly up against handmade barns. The original farmhouse with wooden shingles and three chimneys that once sheltered six previous generations of Orrin's family is appropriately painted white. And, yes, this family has the perfect picket fence.

Jackie and Orrin Pierson

The Pierson Bicentennial Farm is one of the oldest in Orange County. Jackie and Orrin's four children—Lorna, Rachel, Carolyn, and Rita—will be the eighth generation to farm the remaining three hundred of the original one thousand acres of this family's land. Today, they raise pastured grass-fed beef, dairy replacement heifers (in case someone's accidentally jumped over the moon), ornate flowers grown in their extensive greenhouses, a pick-your-own pumpkin patch, and a saw-your-own Christmas tree forest. Jackie's gift shop is absolutely perfect.

"Animals raised in commercial feedlots are confined in large, grassless lots and fed formulated rations comprised mostly of by-product feeds left over from the human food industry, such as potato chip waste, broken cookies, bread crust, and blood meal," Orrin said as we walked through the fields and stopped to watch the calves romp around their mothers. "If I had to do such horrible things to my animals, I wouldn't be able to live with myself. Even the thought of that makes me sick."

For all of her seriousness and extreme dedication as the farmer's wife, Jackie has the ability to make one stop dead in their tracks and burst out in laughter. Her solemn conversational honesty, peppered with unexpected quick wit, is reminiscent of Ellen DeGeneres. She laughs as we race perilously through the fields on the family's golf cart to get to the back pastures. "The hardest part about farming is the lack of a paycheck. We're always constantly chasing what is going to work efficiently out here with Mother Nature and what isn't. Most of the time we've got it figured out, but now we're getting older and we have to revamp everything all over again. We'll have to weld special hooks onto the golf

carts for our canes and walkers. Metal handrails are expensive. We're a little nervous to put those out in the pastures with the electric fencing."

In Orrin's great-great-great-great-grandfather's time, families would have gone into sheer panic to not bear sons to help carry on the family farm. When I asked Orrin about the future of Pierson's Natural Beef Farm, he threw back his head and laughed. "Oh man! Do not even try to tell my girls what they can or cannot accomplish. They're more than ready!" Having met Jackie, I think the "beautiful and tough as nails" genes must run in the family.

Farmer Orrin's Savory Adobo Chili

The question of whether beans "belong" in chili has been a matter of contention among chili cooks for a long time. It is likely that in many poorer areas of San Antonio and other places associated with the origins of chili, beans were used rather than meat, or in addition to meat. American frontier settlers ate chili that consisted of dried beef, suet, dried peppers, and sea salt. These ingredients were mashed together, pounded into a slurry, and dried into bricks. The recipe? Add a brick of chili to a pot of boiling water. (eewwww!)

SERVES 6

3 pounds stew beef, trimmed of excess fat and cut into ½-inch pieces
Sea salt and freshly ground pepper
3 tablespoons olive oil, divided
2 medium onions, finely chopped
3 tablespoons chili powder
2 tablespoons unsweetened cocoa powder
2 cloves garlic, minced
1 tablespoon ground cumin
1 teaspoon dried oregano
2 cups Hemlock Hill Beef Stock *(see p. 215)*
3 tablespoons minced canned chipotle chiles in adobo sauce
1 tablespoon packed brown sugar
1 tablespoon yellow cornmeal
⅓ cup chopped fresh cilantro
Sea salt and pepper to taste
Diced avocado, onion, shredded cheese, and/or sour cream for garnish

Rinse the beef thoroughly and pat dry with paper towels. Salt and pepper beef liberally. In a large, flameproof casserole dish or dutch oven, heat 1 tablespoon olive oil over medium-high heat. Brown beef in batches for 4 to 6 minutes, turning often and removing to a large plate as each batch browns; add more oil to the pot as necessary.

Pour off all but 1 tablespoon of the fat remaining in the pot. Reduce heat to medium. Add onions and cook, stirring often, for 5 to 7 minutes or until onions are softened but not browned. Add chili powder, cocoa powder, garlic, cumin, and oregano. Cook, stirring constantly, for 1 minute.

Add beef stock and bring to a boil over high heat, stirring to scrape up any browned bits from the bottom of the pot. Return beef to the pot along with any juices that have accumulated on the plate. Stir in chipotles and sugar. Reduce heat to medium-low and simmer, uncovered, for about 1½ hours, stirring occasionally, until beef is very tender.

Pour cornmeal over the surface of the chili and stir until well combined. Bring to a boil over medium-high heat, stirring constantly, until thickened slightly. Remove the pot from heat and add cilantro. Season to taste with sea salt and pepper. Serve with your choice of garnishes.

Jackie's Asparagus Frittata

In short, a frittata is an open-faced omelet that is baked in the oven. A frittata is faster and easier to make than a typical omelet and is less-caloric than its older cousin, the cheese-filled quiche.

SERVES 2-4

4 tablespoons unsalted butter
1 medium shallot, minced
1 tablespoon minced fresh mint leaves
2 tablespoons minced fresh parsley leaves
⅓ pound asparagus, cut into 1-inch pieces and blanched in boiling, salted water for 1 minute
5 tablespoons fresh goat chèvre, divided
½ teaspoon sea salt
¼ teaspoon pepper
6 large eggs, lightly beaten
Sea salt and pepper to taste

Adjust the oven rack to the upper-middle position and preheat the oven to 350°F.

Melt the butter in a 10-inch ovenproof skillet over medium heat. Sauté the shallot until softened, about 4 minutes. Add the mint, parsley, and asparagus and toss to coat with the butter. Arrange the asparagus in a single layer. With a fork, whisk 3 tablespoons goat cheese, sea salt, and pepper into the lightly beaten eggs. Pour the egg mixture into the skillet and cook over medium heat until the bottom is firm.

Crumble the remaining goat cheese over the top of the egg mixture. Place the skillet into the oven and bake until the top is just barely set and the center is dry to the touch, about 8 minutes.

Run a spatula around the skillet edge to loosen the frittata. Invert the frittata onto a serving plate, cut into wedges and serve warm. Season to taste with sea salt and pepper.

W. Rogowski Farm

379 Glenwood Rd., Pine Island, NY 10969
(845) 258-4574 | www.rogowskifarm.com

Pine Island, home of the W. Rogowski Farm, was so brown it looked like Death Valley. Not long ago, there were black dirt fields that issued forth green edibles from every nook, cranny, and pore. Farmer Cheryl Rogowski grew food on 150 acres; fed more than five hundred people through her community supported agriculture (CSA) program; sold produce through five farmers' markets; served an on-the-farm-raised breakfast every weekend; fed hundreds of hungry visitors throughout the Hudson and Delaware Valleys at her monthly gourmet supper club, catered events, and private dining services both on and off the farm that showcased her famous diverse and unusual produce; and had her own line of Black Dirt Gourmet prepared and baked goods. Recently, readers of the *Times Herald-Record* voted the W. Rogowski Farm one of the best farm-to-table experiences in all of Orange County.

In 2011 Cheryl Rogowski lost 99 percent of her entire farm to Hurricane Irene and Tropical Storm Lee. All of it is gone. Most people think of losing a farm as similar to losing a house in a flood. Things are inside that house. Things get wet, moldy, and destroyed. And then one collects insurance payments, rebuilds the external structure with new things, and buys new things to put back inside. But to lose a farm is definitely not the same.

If you lose your farm to a flood, you cannot collect all of your dirt and throw it to the curb for trash pickup as one would a wet sofa. Since floodwater contains sewage, chemicals, heavy metals, pathogenic microorganisms, and other contaminates, federally mandated regulations require that all crops exposed to any floodwater be destroyed before the food enters the human food supply chain and disposed of at great distances away from future food production sites. Cheryl not only lost all of her farm investments this year, she lost her income. She will still have to invest in soil remediation, soil testing, and planting of cover crops to clean up the floodwater's polluting damage. In short, farming expenses don't stop simply because one is not currently in production or not generating any income. "The electric meter keeps running, the phone bills keep coming. And right smack dab in the middle

Cheryl Rogowski

of the hurricane, the tax bill arrived. It was a bit of insult to injury," Cheryl said.

I met Cheryl three weeks after the storm, and she said, "A few days ago we were still covered by about five feet of water. Now that it's receded, I wish it was back under water again. Just look! It's hard to see that it's actually all gone. This is intense."

To be a farmer means to invest time and money and work eighteen-hour days for at least seven months straight without a paycheck. Then the harvest comes due. If you're lucky, no natural disasters have happened, the federal agricultural policies are still allowing a small sliver of market space in which to generate an income, and there are buyers and local food outlets through which to distribute your products and hopefully realize some profit after all of the time, money, energy, and expenses that have been invested during the agricultural year. And farmers like Cheryl save their seeds. What they planted this year and what they were hoping to plant next year are all vegetables with a significant amount of family history. The value of a squash is more than just that of a simple plant.

And, don't forget that farmers also employ other people. "I had fourteen farmers this summer," she said. "They watched this farm start to go under water and they immediately started looking for work elsewhere. A few of them were lucky enough to find other jobs. I can't provide them employment this year. I'm now relying on volunteers.

"We have to excavate and clean every one of our drainage ditches. I know that every one of our culverts must be lifted—intact—with heavy equipment and cranes and reset in their proper locations so that we have adequate drainage on

the farm," Cheryl points out. "We have to crawl on our hands and knees and start pulling out the acres of plastic that were underneath the tomato plants and watermelons. It's buried under three feet of mud. We have to pay for all of this. And with what? I just lost my paycheck for this year."

On the second day of Hurricane Irene, Cheryl wrote on her blog, "How do you describe what is happening? How do you describe what has happened? How do you describe what is to come? It starts with the warnings: 'There is going to be a storm, a big one. It's coming.' At first you think 'okay, we've had storms before. I've lived through hurricanes here before. We'll hang on and deal with it like we always do.' The warnings escalate. 'This is a big one, category 1, no 2, no, category 3!' And the projected path has us in the line of fire. I calculate how much I think we can sell, how many deliveries we have scheduled, what can we store the best. I strip all the tomatoes, cut greens, and pick the peas. I check the weather reports. This is not good. How far out can we harvest and hold product? Work faster, harder. How much can we pull? Stuff the refrigeration units! The sky darkens. The heavens open up and there's no stopping it. I keep taking pictures so I can share what once was."

New York governor Andrew Cuomo estimated that storms Irene and Lee caused the destruction of upward of $40 to $45 million dollars in crops. Some estimates are as high as $85 million in crops lost. These include raw tomatoes, peppers, beans, eggplants, and pumpkins. What aren't mentioned in these numbers are collapsed drainage ditches, damage to equipment such as plows and tractors, buildings, farm stands, farmhouses, fences, and everything else. Also uncounted are dead animals, damaged maple trees, destroyed beehives, or the value-added products such as honey, eggs, salsas, jams, jellies, and prepared foods.

Federal crop insurance was created in the 1930s Dust Bowl days to help farmers survive the ravages of nature. In the mid-1990s, Congress privatized the program. The US Treasury still guarantees the riskiest policies using our taxpayer dollars but turned over the selling and servicing of policies to private insurance companies. According to campaign finance reports, over the last twelve years political action committees for the crop insurance industry have donated millions to national political campaigns. In 2005, National Public Radio (NPR) broke a nationwide story indicating that, as a result of spotty oversight, the crop insurance program controlled by private industry, and paid for by the government, is ripe for fraud. In 2011, the *Los Angeles Times* reported that farm insurance fraud is still cheating taxpayers out of millions and further asserted that farmers, along with "their crooked insurance

agents and claims adjusters in the private industry" are fraudulently claiming that weather or insects are destroying their crops in order to cash in on a government-backed insurance program that is bilking our nation out of millions of taxpayer dollars every year. After Hurricane Irene and Tropical Storm Lee, Mitt Romney, like all of the 2012 Republican presidential candidates, called federal disaster relief for flood victims like Cheryl "immoral" and said it "makes no sense at all." The federal government nearly shut down in September because of partisan disputes on funding for disaster assistance.

After the hurricanes, small farms like the W. Rogowski Farm that grow vegetables for our farmers' markets and for our CSA programs found themselves silenced under the merciless thumb of national corporate agriculture's best interest and in the middle of ridiculous congressional politicking for election status. Since Cheryl's farm does not produce commodity crops, it therefore does not qualify for adequate crop insurance coverage. She has to wait in the federal disaster relief breadlines. But industrial farms that grow wheat, corn, soybeans, rice, or cotton receive federal agricultural subsidies and also qualify for adequate crop insurance coverage—a double tax-paid bonus for which they can thank their lobbyists. Industrial-scale farms that produce acres of a single homogenized crop, such as, say, tomatoes for the Campbell's Soup Company, qualify for crop insurance as well, for which they, too, can thank their lobbyists. Multi-crop-growing small farms, like all of the farms in this book—the farms that we know at the farmers' markets, the farms that produce apples and lettuce and peas and broccoli and garlic in rotational planting on smaller plot sizes—do NOT qualify for large-scale crop protection coverage, do not receive any agricultural subsidies, and do not have any lobbyists advocating on their behalf.

According to the US Department of Agriculture, "Farming is a high-risk business." In more ways than one. Cheryl has certainly learned the harsh reality of that statement.

But in her response to disaster, Cheryl Rogowski also reminds me of Laura Ingalls Wilder. Perhaps it is the pragmatic way she is enduring such a devastating tragedy. Perhaps it is her long gray hair swept back in a bun, or the way she uses the words *homestead* and *balance* and *survival.* As we spoke, I was keenly aware that without our commercialized food system, hurricanes like these that obliterated the local food supply in Orange County would have resulted in deaths from starvation over the winter months, as occurred in Wilder's lifetime.

We are 100 percent dependent on Mother Nature for our food supply. Yet it is really easy to forget this basic fact when, after a hurricane,

one can go to a supermarket and buy food—any kind of food—at the same time a neighbor's food-producing fields are starting to flood next door. No wonder Cheryl is troubled and puzzled. As I surveyed the lifeless and birdless landscape, Cheryl asked, "Am I going to lose market share at the farmers' markets? How many of my customers, particularly my CSA customers, will hang in there with me, knowing that the basic premise of a CSA is to share in the farm's risk? This is an extreme situation, but crop loss happens and must be faced. Perhaps all of my customers will quit and go back to the grocery store where it's easier to forget how hard farming really is and make all of this for naught."

I have dedicated this book to Cheryl, the W. Rogowski Farm, and the many other farms just like hers that I was not able to include in this limited book. When you read this, know that Cheryl still needs your help. She can also point you to her neighbors who also fall within the same out-of-balance, ever-widening gap between megasize corporate farming and small family farmers who have no one there to help them but us. Throughout our interview, we stood in the sun on the deck and surveyed the endless acres of mud and ruin. This is the most devastating natural disaster that she has ever experienced in her lifetime, yet she was kind, honest, gentle, realistic, and welcoming. I know that if this farm survives, Cheryl

will not be able to generate full income potential for at least three more years. Everything is gone and must be rebuilt. Left unspoken was our common fear that she might not financially endure that long. But, she's already back out in the fields, alone. She is going to try. I know that I do not have the capacity to be so brave.

Cheryl's Orange Cucumber Soup

Cooked cucumbers are quite tasty and most other cultures and cuisines cook cucumbers often as part of their culinary repertoire. If you're feeling adventuresome, give this a try.

SERVES 4

1 tablespoon olive oil
1 orange, cut crosswise into paper-thin slices, seeds removed
½ cup diced sweet onion
1 cup diced yellow or red bell pepper
3 tablespoons chopped fresh parsley
2 medium cucumbers, peeled and cut into ½-inch pieces
3 cups stock of your choice (see p. ix)
¼ cup heavy cream
¼ cup sour cream
Sea salt and pepper to taste
3 tablespoons minced fresh chives

Heat the olive oil in a large saucepan over medium heat. Sauté the orange rings until golden on both sides and all of the juice has been released. Remove the oranges and set aside for garnish.

Add the onions, bell peppers, and parsley to the saucepan and sauté for 4 or 5 minutes until the onions are soft, but not brown. Add the cucumbers and cook for 1 minute. Add the stock, cream, and sour cream. Increase the heat to high and bring to a quick boil. Reduce the heat and simmer on low for 5 minutes or until the cucumbers are very soft. Puree the soup in a blender or using an immersion blender until creamy and smooth. Season to taste with sea salt and pepper.

To serve, ladle the soup into serving bowls. Garnish with seared orange pieces and fresh chives.

Rogowski Ruby Red Vegetable Stock

Vegetable stock is simple to make, and yet, for some reason, most of us struggle with it—often ending up with some kind of pithy-tasting, off-colored water instead of an actual stock. Add this recipe to your Rolodex for perfection every time.

MAKES 1 QUART

3 tablespoons unsalted butter
2 medium onions, coarsely chopped

8 large shallots, sliced thin
3 roasted red peppers in oil, oil removed with paper towels, minced
8 cups hot water
2 stalks celery, coarsely chopped
4 carrots, peeled and coarsely chopped
1 small head cauliflower, cored and chopped fine
1 red bell pepper, seeded and coarsely chopped
1 yellow bell pepper, seeded and coarsely chopped
10 fresh thyme sprigs
2 bay leaves
1 teaspoon black peppercorns
Sea salt and pepper to taste
2 teaspoons mirin or rice vinegar

Melt butter in a large stockpot and sauté onions and shallots over medium heat until browned and beginning to caramelize. Add the roasted red peppers and sauté for 1 minute more, stirring constantly.

Add water, celery, carrots, cauliflower, red and yellow peppers, thyme sprigs, bay leaves, and peppercorns. Bring to a boil over high heat. Reduce the heat to low and simmer gently for about 2 hours until reduced by half.

Strain the vegetable stock through a fine mesh sieve. Do not press on the solids. Season to taste with sea salt, pepper, and mirin. Use immediately or refrigerate for up to 4 days.

Orange County Post-Hurricane Healing Pumpkin Punch

SERVES 6-8

1 pumpkin, lid cut out, seeds and pulp removed
3 (11.2-ounce) bottles Pumpkin Ale, cold
3 (11.2-ounce) bottles hard apple cider, cold
2 cinnamon sticks
5 whole cloves
Toasted pumpkin seeds

Carve the pumpkin to your heart's delight (or not, depending on your mood.) Place a glass bowl inside the hollowed-out pumpkin.

In the bowl, combine the cold Pumpkin Ale, hard apple cider, cinnamon sticks, and whole cloves. Stir to combine.

Serve with toasted pumpkin seeds to friends.

Chapter 5

Putnam County

Dancing Girl Bees & the Midnight Farmer

37 Spruce Mountain Dr., Putnam Valley, NY 10579
(845) 528-2157

Dennis and Minnie grew up in the Bronx. They met and married. They maintained good careers, are good citizens, and no doubt drive the speed limit—well, I think Dennis does anyway. Now, they carry a new title: farmers.

Dennis Velez, former senior city property manager for Cushman & Wakefield Commercial Real Estate Brokers, spent his early years with his grandmother in Puerto Rico. If Dennis had been born in America, we'd say that he "grew up on a farm," but running around with his *abuelita*'s chickens and pigs and goats and eating food grown in the family's vegetable garden is a normal way of life on a tropical island that sanctions no specialized elite title for producing what one needs.

Moving from rural Puerto Rico to the urban Bronx, Dennis started to realize that the agrarian knowledge we should have—did have—got lost somewhere along the way. Time passed, as it invariably does, rolling Dennis forward into

Minnie Santos and Dennis Velez

POTTING
SHED
Grandchildren complete
life's circle of love.
BRAND
PUMPKIN

happy marriage with a delectably curly-haired, pint-size woman named Minerva, and on into his middle years, consumed with the exhausting hours of a successful career. And then, the mortgage crisis happened. His employer downsized Dennis out onto the streets of New York City to fend for himself. "We were dreaming of a more self-sufficient lifestyle, but I was filled with terror," Dennis said. "Going to work was really all I knew." Through all of these personal upheavals, Dennis and Minnie followed their gut instincts, and the Dancing Girl Bees & the Midnight Farmer farm was created on their property in Putnam Valley. I found listening to these two nouveau farmers tell about their early misadventures—in the typical back-and-forth storytelling fashion of happy couples—hysterically funny.

At Dancing Girl & the Midnight Farmer, Dennis and Minnie are building camps, hiking trails, and summer retreats on the back side of their forested property. In the meantime they operate a large apiary and provide wild honey to anyone who wants some. Their chickens produce eggs for Minnie's patients at her medical practice. Their frog pond and waterfall will keep you captivated for hours. The fruits of their gardens feed anyone who is sick or well or just plain hungry.

"It took Minnie and me many years to learn the lost skills so natural to my grandmother: husbandry, raising chickens, growing food, dealing with animal illnesses, preparing our farm for damaging storms, and more." And then, Dennis got sick. Ten surgeries were needed to repair and replace whole pieces of his heart. "Farming kept me alive," he said. "I was a caretaker. I was responsible for other things. This gave me the mental and emotional strength to work through the depression that accompanies the endless hours of bed rest, pain, and the overwhelming fear of dying."

Minnie is an internal medicine physician who just finished her two-year fellowship in integrative medicine with Dr. Andrew Weil, and was named "Top Doctor" by *Westchester Magazine.*

Her journey into farming started around 2006. At the time, this extraordinarily busy doctor was tending to patients, hosting her own TV show, and serving as medical director for Northern Westchester Hospital and as an associate professor at New York Medical College. "I was a typical type A personality," she said. "I'd get up early in the morning, throw myself into the shower, slam down a cup of coffee, and race out the door to go to work. It's funny how life happens. I actually just always wanted chickens. I tell people all the time that chickens changed my life."

Chickens, as you know, are early risers. To ensure their safety at night and their ability to roam unattended as "cage free," Minnie was up early every day to release the hens and to gather eggs. Now, however, instead of rushing everywhere at breakneck speed, she stands outside in slippers and an overcoat with her hens, sipping coffee while the sun rises. "I suddenly realized that there is a whole other world that was here, all around me. It was beautifully quiet and calm. The birds and bees and wild animals were waking up and going to work. Nothing in working nature is hectic or frantic; there is an overwhelming serenity." She suddenly realized the health and medical value of calming, quiet time. Neither her medical practice nor her personal life has since been the same.

Dennis and Minnie's farm is the very embodiment of perspicacity. Brightly colored arbors are vehicles for the vines of bean and peas. Red tables and a plethora of mismatched decorative ornaments accompany vibrant deck chairs around the central slate-based fire pit. The pluming weeds of goldenrod wave proudly at the base of the steep slopes as an exquisite all-you-can-eat bee buffet.

As for their name, the Dancing Bees & the Midnight Farmer? Minnie laughs. "She used to be a dancer when she was young," Dennis pokes. "And he absolutely hates to wake up early," Minnie retorts. After completing most of the chores before Dennis even got out of bed, Minnie demanded, "What are you? A midnight farmer?" And so, the name of their farm represents their two unique personalities, backgrounds, histories, needs, and personal desires. It's perfectly symbiotic.

Dancing Girl Caramelized Vanilla Tart

What would a cookbook be without at least one bottom-widening dessert recipe? This is pudding for grown-ups and can be further customized by adding flavoring to the caramel such as 1 tablespoon of fresh ginger or 1 teaspoon of cinnamon or whatever strikes your fancy. When I grow up, I want to eat this.

SERVES 12

Zan's Honeyed Tart Crust (see p. 36)

Vanilla Cream

1 vanilla bean (or 1 tablespoon vanilla extract)
1 cup whipping cream
1 cup whole milk
3 tablespoons honey, divided
2 egg yolks
3 tablespoons cornstarch

Caramel Sauce

⅓ cup water
¾ cup granulated sugar
⅓ cup whipping cream
⅓ cup unsalted butter, cut into pieces

Prepare and bake Zan's Honeyed Tart Crust. Set aside to cool.

To make the vanilla cream, cut the vanilla bean in half lengthwise and scrape seeds out with the back of a paring knife. Reserve the seeds. In a small saucepan, heat the cream, milk, half of the honey, and the vanilla bean pod (or extract) over medium heat until just simmering. In a separate bowl, whisk together remaining honey, egg yolks, and cornstarch. Slowly pour in the hot milk mixture while whisking constantly. Return to the saucepan and heat, whisking very gently, for about 5 minutes over medium-low heat until mixture beings to boil and thicken. Remove the vanilla bean pod and whisk in the seeds. Transfer mixture to a large bowl and cover with plastic wrap pushed directly into the vanilla cream to prevent contact with air. Refrigerate for at least 2 hours to allow the custard to cool and set.

To make the caramel sauce, add ⅓ cup water to the sugar in a heavy saucepan and bring to a boil. Boil, stirring constantly, for about 3 minutes, swirling the pan gently to distribute color, or until it is dark amber. Remove from the heat. Slowly add the cream and stir until smooth and the bubbles subside. Whisk in the butter until melted and smooth. Remove from heat and let the caramel cool, stirring occasionally for the next 15 minutes or until it starts to thicken.

Pour all but 4 tablespoons of the caramel into the baked tart shell. Set aside until cool and firm. Spoon the vanilla cream onto the caramel layer and spread evenly. Refrigerate at least 1 hour before serving. Drizzle the remaining caramel sauce across the top before cutting.

Note: If you're feeling lazy, skip the tart crust and serve vanilla cream with a small scoop of your favorite ice cream topped with the caramel sauce.

Midnight Farmer Seared Deviled Eggs

This is a variation on an old French recipe—standard fare for hungry travelers, guests, and friends. These eggs are perfect munchies all by themselves or perfectly yummy atop any salad. Enjoy!

SERVES 6

6 large eggs
1 teaspoon finely minced ginger
7 tablespoons olive oil, divided
1 teaspoon finely minced shallots
2 tablespoons minced fresh tarragon leaves
3 tablespoons heavy cream
½ teaspoon sea salt, divided
½ teaspoon freshly ground black pepper, divided
1 tablespoon white wine or water
Sea salt and pepper to taste

Put the eggs in a small saucepan in a single layer. Cover with enough cold water to come 3 inches above the eggs. Bring to a rapid boil. Immediately remove from the heat and cover the pot. Allow to rest, undisturbed, for 12 minutes. Drain off the water, and shake the eggs in the saucepan to crack the shells. Immediately immerse the eggs in cold water and ice to chill rapidly for 15 minutes.

In a large sauté pan, sauté the ginger in the olive oil until crispy and fragrant. Remove the ginger to a large bowl and reserve half of the olive oil for the dressing.

Shell the eggs under cold running water and split them lengthwise. Remove the yolks carefully and put them in a bowl with the sautéed ginger. Add shallots, tarragon, cream, and half the sea salt and pepper. Crush with a fork to create a coarse paste.

Spoon the mixture back into the hollows of the egg whites, reserving 2 to 3 tablespoons of the filling to use in the dressing.

Reheat the gingered olive oil in the sauté pan. Place the eggs in the skillet stuffed side down. Cook over medium heat for 2 to 3 minutes, until the eggs are beautifully browned on the stuffed side. Remove and arrange the eggs, stuffed side up, on a platter.

For the dressing, mix the reserved gingered olive oil, the reserved egg stuffing, wine, and the remaining sea salt and pepper. Whisk thoroughly until combined. Coat the warm eggs with the dressing and serve at room temperature. Season to taste with sea salt and pepper.

RYDER FARM
GREENS
FLOWERS
POTTED PLANTS
CSA
TOMATOES
HERBS

Ryder Farm

400 Starr Ridge Rd., Brewster, NY 10509
(845) 279-4161 | www.ryderfarmorganic.com

Betsey Ryder has wonderfully arched rosy cheeks, long blond hair, and the deeply mellow voice of Joan Crawford on *Lux Radio Theater*.

In her blue-gingham dress, she talks with her husband, John, about running the Ryder Farm, which has been in her family since 1795 in Brewster, New York. All 128 acres on Peach Lake are managed by interns from around the United States through WWOOF (World Wide Opportunities on Organic Farms), NCAT's sustainable agriculture project, and MESA (Multinational Education for Sustainable Agriculture).

WWOOF is an exchange program. In return for volunteer help, WWOOF hosts like the Ryders offer food, accommodation, housing, and opportunities to learn about organic farming. The National Center for Appropriate Technology (NCAT) provides internship listings as a service to the sustainable agriculture project of the United States Department of Agriculture and the sustainable agriculture community at large. MESA's mission is to advance a new generation of agrarian leaders and to link innovations in sustainable agriculture with ancestral knowledge to promote land stewardship, place-appropriate production, and cultural awareness.

I doubt that Betsey, a spritely fifty-something, would enjoy being labeled "ancestral." Still, Ryder Farm's eight interns gather together from Thailand, Ecuador, Bangladesh, Arizona, Missouri, Maine, the Bronx, and Pittsburgh, Pennsylvania, to produce and harvest Betsey's vegetables in Putnam County using techniques that combine the modern with the tried-and-true. "Not only is it fundamentally important to have young people involved in agriculture," Betsey says, "but it's an integral part of our mission. We are not only growing vegetables, but growing the future of farming." Applications to work on the Ryder Farm come from Fulbright scholars, graduates of Stony Brook University, doctoral candidates from Columbia University, and medical students from all over the world.

Betsey Ryder

Betsey is the ninth generation of her family to operate the Ryder Farm. A tomboy growing up on Long Island, she traveled with her parents back to the family farm in Brewster for the farm's annual meeting every Fourth of July. Over the years, she came to know her family's farm. In the 1960s and into the '70s the Ryder Farm suffered tremendous hardship because of exponentially rising taxes. They took shelter in New York's Agricultural and Markets Protection Law, article 25-AA, a tax-protection opportunity for farmers that paved the way for Betsey to be able to afford to continue farming operations on her family's land beginning in 1978. In 2010, they found themselves again in great need for similar assistance from the state of New York when Betsey's farmland taxes rose 33 percent—an additional $7,500—in less than one year.

Eight acres of Ryder Farm produces organic vegetables for their CSA members, their farm stand, and the Union Square farmers' market. In 1977, they were the first organic farm to sell produce at Union Square Greenmarket in Manhattan, signifying the agricultural transition that this farm made from dairy farming to pesticide-free vegetable cropping. This property has large acres of tilled fields, house gardens, rows upon rows of perennial and annual flowers for the custom-order flower business, greenhouses, walk-in coolers and storage facilities, and painted delivery trucks to move Ryder Farm's products into New York City. And yet, Betsey works tirelessly to make her food-growing space also aesthetically pleasing. Ryder Farm Cottage Industries looks and feels less like the full-scale mega farming operation that it actually is, and more like a cozily comfortable backyard.

An intricately carved gazebo rests in the middle of the farm field. Betsey's right in saying, "It is a beautiful thing to be able to relax for a moment, rest in the shade, and look out over the farm and enjoy the fruits of one's labor in such a romantic way." Betsey calls the Ryder Farm a "true gentleman's farm." After the Great Depression, this gazebo once overlooked the clay tennis court festooned with roses on every pillar. Listening to Betsey talk about farming and food and family, you can almost see the ghosts of lovely ladies in sweeping embroidered skirts and cinched corsets. Her ancestors twirled their parasols and drank ice tea on Sunday afternoons, while watching the boys play tennis on their day of rest.

Ryder Farm Fresh Oysters with Peach Mignonette Sauce

Peaches, oysters, and vodka . . . enough said. If you don't like oysters, try this sauce on grilled hamburgers or grilled vegetables.

SERVES 4

2 very ripe peaches, peeled, pitted, and minced
¼ cup finely minced shallots
½ teaspoon sea salt
4 tablespoons vodka
16 bluepoint oysters on the half shell

Over a large bowl, mash the diced peaches through a fine mesh sieve to extract juice and mash the fruit.

Stir together peach pulp, shallots, sea salt, and vodka. Allow to sit for 40 minutes at room temperature.

Top each freshly shucked oyster with 1 tablespoon of peach mignonette sauce and serve chilled.

Betsey's Tomato & Mozzarella Salad

There are lots of versions of tomato salad and tomato and mozzarella salads out there. This lemon-cumin dressing sets this recipe apart. Too, even if it isn't tomato season, keep the dressing as part of your winter repertoire as it's perfect with hearty greens. Or, if you're really adventurous, poach tilapia in the dressing and then serve atop the tomatoes and mozzarella balls!

SERVES 4

8 ounces small, fresh mozzarella balls, halved
4 cups cherry tomatoes, all colors, halved
1 teaspoon freshly grated orange rind
⅔ cup olive oil
⅓ cup lemon juice
1 teaspoon sea salt
1 teaspoon freshly ground black pepper
1 teaspoon ground cumin
Sea salt and pepper to taste
¼ cup minced fresh parsley

If using small mozzarella balls, halve them into ¼-inch pieces, or use a melon baller to cut rounds out of a larger piece of fresh mozzarella.

Toss mozzarella, tomatoes, and orange rind together in a large bowl.

In a glass jar, combine olive oil, lemon juice, sea salt, pepper, and cumin. Close the container and shake vigorously for about 1 minute, or until all ingredients are well blended.

Pour half of the dressing over the tomatoes and mozzarella. Toss to combine, cover, and set aside. Just prior to serving, toss the tomatoes in the dressing again. Feel free to add more dressing if you'd like. Season to taste with sea salt and pepper. Toss with fresh parsley and serve.

Gentleman's Farm Savory French Toast

French toast doesn't necessarily have to be a super sweet breakfast that sends your blood sugar levels through the roof. Try a savory version that is, actually, perfectly wonderful for lunch or dinner!

SERVES 4

3 large eggs
2 tablespoons milk
1 teaspoon sea salt
½ teaspoon pepper
1 loaf good-quality raisin bread, unsliced
2 tablespoons honey mustard
8 ounces smoked ham, shaved
5 ounces soft white farmers' cheese, sliced thin
4 tablespoons unsalted butter, divided

In a square baking dish, lightly beat the eggs, milk, sea salt, and pepper. Cut 8 ¾-inch slices of raisin bread. Spread 4 slices with honey mustard, top with ham, cheese, and a second slice of bread. Press firmly.

Dip the 4 sandwiches in the egg mixture to coat each side, making sure that all of the egg is absorbed.

In a large nonstick skillet, melt 2 tablespoons butter and fry one side of each sandwich until golden. Flip the sandwiches, add the remaining 2 tablespoons of butter, and cook until golden, about 5 minutes more.

Slice in half and serve warm.

Tilly Foster Farm Museum

100 Route 312, Brewster, NY 10509
(845) 228-4265 | www.tillyfosterfarm.org

I met Ann Fanizzi on an oppressively humid, foggy morning in August. I, like most photographers, cringed from the heavy, impenetrable morning moisture. At first. Then, when I looked carefully, there was magic. Horses glowed against the fog like long-forgotten ghosts. Twin headlights sliced from side-to-side up the winding driveway. A farmer in muck boots and overalls delivered his gift, Mystery the rooster, in a cat carrier and then disappeared again with his red taillights. The fog obliterated all sight of the frenetically busy Route 312 upon which the Tilly Foster Farm resides, and it forced the otherwise booming crawl of endless traffic instead to echo demurely somewhere in the overhead trees.

Ann is the interim acting director of the Tilly Foster Farm Museum. This farm was purchased in 2002 with East of Hudson Water Quality Funds and with help from the Putnam County Coalition to Preserve Open Space—a successful partnership between private citizens, advocacy groups, and county and state government officials. All 199 acres were rescued from foreclosure.

The nonprofit Tilly Foster Farm Museum is now a farming preservation and agricultural education center under the supervision of the Whipple Foundation and the Society for the Preservation of Putnam County Antiquities & Greenways. Their goals are to keep the Tilly Foster Farm free and open to the public, to make the farm as self-sufficient as possible, to educate the public about farming's local history, and to build a world-class farm museum featuring rare and endangered early American farm animals.

The extensive buildings, barns, and houses have all been preserved and restored. A delightful display of antique tractors and Model T Fords now occupies the former milking parlor. One of the many owners, the Benedict family—like

Ann Fanizzi

Putnam County
Tilly Foster
Farm
Conservation
Area

every other New York dairy operation—dismantled its dairy herds in the 1950s and early '60s.

Centralized dairy farming primarily developed around villages and cities where urban residents were unable to have cows of their own. Dairy farms surrounded New York City in concentric rings, and Putnam County's dairy farmers were no exception. They would hand-milk their herds twice a day, fill barrels and pails with milk, and transport their products into the city. Since an individual could not be expected to milk more than a dozen cows per day, New York's dairy farms were small and numerous. In fact, refrigeration equipment was first used in the United States in the early nineteenth century as part of the dairy farming operations to keep milk cool prior to its transport to market.

In the 1950s direct-expansion cooling for large milk parlors and storage tanks, combined with the use of electricity for milking purposes and the invention of vacuum bucket milking machines, forced many small dairy operations out of business, and allowed the survivors to increase their herd sizes a hundredfold. As production went up, prices went down. In 1949 as World War II wartime price controls ended, the federal subsidies for milk became federal dairy policy, upsetting the balance between the supply of milk and the consumer demands for milk. At that time, small family dairy farms could no longer generate profit and disappeared.

Having been squeezed out of the dairy business, the Benedicts' Dairy Farm became the Tilly Foster Farm and embarked on breeding thoroughbred horses for the Kentucky Derby. Their claim to fame was the breeding and racing of multiple third-place finishers at Churchill

Downs. Mr. Benedict became president of the New York State Thoroughbred Horses Association. He constructed event space for hosting association members and public officials, throwing glittering celebrity receptions and teaching workshops and conferences on horse racing. Many of these buildings remain preserved today.

The Tilly Foster Farm Museum is an impressive sight. Modern solar power panels and small wind turbines stand in contrast to the odd and extremely rare Randall lineback cattle, mammoth jackstock donkeys, American blue rabbits, Pilgrim geese, Indian runner ducks, Narragansett turkeys, Delaware chickens, and a pair of very vociferous black American guinea hogs.

This preserved space is a perfect blend of old and new. Awestruck schoolchildren dance through ornately carved historical buildings. Fathers and grandfathers hang out in the milking parlor, rubbing their hands over antique trucks and tractors and telling stories. Mothers watch the rare and perfectly white American cream draft horse nuzzle grass in the pasture and remember their own childhood fantasies.

Tilly Foster Bacon Potato Soup

Hmmm, comfort food. If its snowing, and you're cold, and you have about 20 minutes worth of energy left with which to cook dinner, then this is the recipe for you. Snuggle up for a 30-minutes-or-less meal-in-a-pot dinner that'll ease your chilled fingers and aching back.

SERVES 6

½ cup thinly sliced double-smoked bacon
1 cup sliced leeks
2 tablespoons finely minced lemon zest
¼ teaspoon nutmeg
3 cups peeled, diced potatoes
½ teaspoon sea salt
½ teaspoon pepper
1 cup dry white wine
3 cups stock of your choice (see p. ix)
½ cup sour cream
2 tablespoons freshly squeezed lemon juice
Sea salt and pepper to taste
¼ cup minced chives

In a large, heavy stock pot, cook bacon over medium-high heat, stirring often, until crispy, about 6 minutes. Using a slotted spoon, remove the bacon and reserve for garnish. Drain all but 2 tablespoons of fat from the pan. Add the leeks and cook until softened. Add the lemon zest and nutmeg and sauté, stirring constantly, for 1 minute more.

Add potatoes, sea salt, pepper, and wine and cook until reduced by half. Pour in stock and reduce heat to maintain a simmer, cooking until the potatoes are very tender, about 15 minutes.

Using a blender or an immersion blender, puree soup until creamy and smooth. Stir in the sour cream and lemon juice. Season to taste with sea salt and pepper.

Serve in warmed soup bowls topped with bacon and chives.

Ann's Spinach Salad with Bacon Queso Dressing

You can serve this salad without the bacon too, by using a bit of roasted eggplant or moutabel. Going one step further, you can sauté any other kind of protein after the bacon—such as chicken. Magically, you'll have a seared Chicken Salad with Spinach and Bacon Dressing. Oh! The possibilities!

SERVES 4

2 strips thick-cut bacon
¾ cup olive oil
¼ cup rice wine vinegar
3 tablespoons whole-milk yogurt
3 tablespoons grated queso fresco, queso blanco, or Romano cheese
¼ teaspoon sea salt, plus additional if necessary
½ teaspoon pepper, plus additional if necessary
6 cups fresh baby spinach
2 cups snow peas, trimmed and cut in half
2 tablespoons minced cilantro
Freshly ground pepper to taste

In a small saucepan, cook bacon until crispy; crumble and set aside to drain on paper towels. Reserve 1 tablespoon bacon fat.

In a blender, combine 1 tablespoon bacon fat, olive oil, vinegar, yogurt, queso fresco, sea salt, and pepper. Blend until smooth. Adjust seasonings if necessary. Stir in crumbled bacon. Pour into a jar and secure lid and refrigerate until ready to use. Shake vigorously before using.

To serve, toss spinach and snow peas together in a large salad bowl. Distribute among plates and top with dressing. Sprinkle with cilantro and a dash of freshly ground pepper. Serve immediately.

THIS WAY TO...
GOATS
PIGS
CHICKENS
SHEEP
COWS
HORSES
BACK SOON
ALPACAS

Chapter 6

Queens County

Las Delicias Patisserie

36-46 37th St., Island City, NY 11101
(214) 417-0044 | www.lasdeliciaspatisserie.com

You may be surprised that, in this book celebrating farmers, I have placed a confectioner. But when it comes to baking, the sum is less than the whole of its parts.

Without preservatives or artificial flavors, each handmade Las Delicias pastry is baked with the finest, purest, most wholesome ingredients sourced directly from local farmers. Debbie Brenner, founder and owner of Las Delicias Patisserie, yearns for everyone to experience the numerous flavors and textures of the real Old World art of baking. All of her products are vegetarian and kosher and are baked with regionally produced organic flour, free-range eggs, milk, and butter. All humanely produced ingredients are procured through fair-trade agreements and come from farms that petition against the unfair exploitation of labor.

Debbie Brenner

Mrs. Brenner is, by her own admission, "a very adventurous person." She had a good job once, working for UPS for more than twenty years. One day this Argentinean, fluent in four languages (Italian, French, English, and Spanish), decided to quit her job, walk away from her pension plan and health insurance and regular paycheck, and start baking. She created Las Delicias Patisserie, and, being one of one-hundred-plus entrepreneurs using the valuable business incubator the Entrepreneur's Space, a centralized shared-business space for small food producers funded by the New York City Economic Development Corporation, Debbie is one of the very first approved bakeries at GrowNYC farmers' markets.

With a regular pastry line as well as a phenomenal gluten-free product line specially formulated for her husband, who suffers with celiac sprue (an intestinal disorder that interferes with digestion and the absorption of nutrients), Las Delicias Patisserie is the only 100 percent locally sourced gluten-free bakery in all of New York and New England. Trading a nine-to-five, corporate

America banking job, Debbie now goes to work at midnight on Thursday and ends her shift at 3 p.m. on Monday afternoon. Everything is cooked fresh daily and brought to a multitude of farmers' markets throughout the city. She's the happiest she's ever been.

As for consumers, the general palate has become accustomed to the difficult-to-rot muffins, scones, cookies, breads, pies, and tarts provided through the commercialization of our food supply. Over time, we've forgotten what real butter rolled into snowy flour actually tastes like. We're now accustomed to the tastes of partially hydrogenated oil instead of butter, high fructose corn syrup instead of brown sugar or maple syrup or sorghum, artificial colors and flavors to make the carrots more orange or the blueberries more blue, benzoates, whitening agents (wheat is brown, but for some reason we demand white), egg and milk powder instead of just plain eggs and milk, and lecithin—a hydrolyzed by-product of either eggs or chicken and sheep brains or fish eggs or fish roe or blood or bile or soybean oil. (By the way, lecithin is used to eliminate foam in water-based paints, to serve as an anti-gumming agent in gasoline, to prevent sugar crystallization in chocolate, preserve the shelf life of your Twinkies, protect yeast cells in dough when frozen, and work as nonstick cooking spray.)

Although we eat lecithin on a daily basis, I suspect it actually tastes good only because extra sweeteners have been added to mask its true flavor. Both the calorie and nutritional content of Debbie's baked goods are widely different from their commercially produced counterparts. She has justified my long-standing irritation with the misleading "baked goods and fats are bad for you" nutritional marketing featured on the covers of every major "you're fat, be skinny" magazine. Confectionaries made with ingredients harvested directly from the farm are healthier because they are higher in vitamins, minerals, fiber, calcium, and trace elements and are lower in sugar. Baked goods made commercially with some ingredients created by men in white lab coats contain nutrient- and fiber-free calories and leave you more nutritionally deficient than before you ate them. The latter is indeed bad for you, but not all flour-based goods are created equally. Everything produced at Las Delicias Patisserie, however, is appropriate for a healthy, balanced diet—both psychologically and physically!

Today, most of our nation's wheat is grown on vast fields and milled in factories in the midwestern wheat belt of the Great Plains. Similar to corn, this grain is bred for uniformity and mass yield. Specialized seeds and specialized pesticides determine the quantity of the wheat crop with little deference paid to flavor or nutrition. According to the agricultural biotech corporation

Monsanto's website, "Our focus is on delivering high-performing, locally adapted varieties across all of the major wheat classes in the United States. Each year, our strong team of renowned breeders and scientists evaluates thousands of wheat candidates to deliver several new, specialized, proprietary varieties that meet the demands of growers and food processors."

In fact, most people don't realize that there is a resurgence of local farmers in New York State producing heirloom wheat for local consumption. A consortium of farmers and bakers called Northeast Organic Wheat is developing local wheat-to-bread systems and fostering local training, transportation, and milling networks to help ease consumers' extraordinary desire to avoid commercially produced flours. The taste, tang, consistency, essence, and aroma of locally grown-and-ground grains makes us understand what we're truly missing by consuming commercial, industrial-farmed grain, the same way the guttural experience of consuming a locally grown, August-ripened local tomato causes us to bemoan the mushy taste of tomatoes from foreign countries, picked green and ripened by gases and shipped thousands of miles to be sold to us in January.

Debbie sources almost all of her ingredients locally. And not just flour: mushrooms, carrots, onions, herbs, berries, and more. Endlessly more. She needs all of these things for her savory quiches and tartlets made using South American recipes that reflect Italian influences and classic French techniques. "I grew up in Argentina," she said. "I grew up with organic before even knowing that 'organic' was anything special. Cows were hanging out and eating grass. Chickens ran around all over the place. We ate food. When I moved to NYC, I remember walking along the Upper West Side. I saw the most beautiful strawberries you have ever seen—ruby red, plump, shiny. You could eat them with your eyes! I learned the hard way that the only thing you can do is eat them with your eyes. You buy them, and they taste like nothing. They're virtual."

Las Delicias Patisserie uses musky-fragranced, locally produced heirloom wheat. The resulting taste and texture of Debbie's products make other foodstuffs created by supermarket flour seem one dimensional, flat, and slightly bitter by comparison. "I want to reconnect people with the pleasure of eating," Debbie says. "People are too focused on quantity and not eating for pleasure." She makes the cookies our grandmothers ate. Until you meet her, I can honestly say that you have no earthly idea what you're missing.

Debbie's Olive Oil–Poached Bread

You probably think that there is a typo in this recipe. Poaching bread? With olive oil? This recipe is a little more labor-intensive than other bread recipes—requiring 4 separate rises. However, the crispy fried shell of the bread requires no additional condiments. Sit back, crunch, and enjoy.

MAKES 1 LARGE LOAF

Starter Sponge

1 cup bread flour
1/4 cup whole wheat flour
3/8 tablespoon instant yeast
1 1/4 teaspoons honey
1 1/3 cups room-temperature water

Flour Mixture

1 3/4 cups bread flour, plus additional as needed
1/2 teaspoon instant yeast
1 1/2 teaspoons sea salt
1 1/2 cups olive oil, divided

To make the sponge, combine the bread flour, whole wheat flour, yeast, honey, and water in a large bowl. Whisk to incorporate air until very smooth, about 2 minutes. The sponge should be the consistency of a thick batter. Scrape down the sides of the bowl and set aside, covered with plastic wrap, while you make the flour mixture.

To make the flour mixture, mix together the bread flour and the instant yeast. Gently scoop it onto the sponge to cover it completely. Cover the bowl tightly with plastic wrap and allow to ferment for 1 to 4 hours at room temperature, or until the sponge starts to bubble through the flour mixture.

With a mixer and dough hook, mix the bread dough on low speed (#2 if using a KitchenAid) for about 1 minute until the flour is moistened enough to form a rough dough. Or, you can hand-knead the dough (at this and each step) until very elastic, smooth, and sticky enough to cling slightly to your fingers, about 15 minutes. Scrape down any bits of dough. Cover the top with plastic wrap and allow the dough to rest for 20 minutes in the mixer bowl.

Sprinkle on the sea salt and knead the dough on medium speed (#4 if using a KitchenAid) for about 7 minutes. The dough should be very elastic, smooth, and sticky enough to cling slightly to your fingers. If it is too sticky, add additional bread flour, a tablespoon at a time.

Scrape the dough into a large bowl and push it down to extract any air. Pour 3/4 cup of olive oil over the dough and turn it so that all sides of the dough and the bowl are covered in oil. Cover the container with plastic wrap and allow the dough to rise in an area where the temperature is 75°–80°F until doubled, usually about 1 hour.

Punch the dough down in the bowl to extract all of the air. While still in the bowl, fold it into thirds, like a letter, and turn the dough upside down in the bowl. Add the remaining 3/4 cup olive oil and swirl to coat all sides of the dough. Cover the container with plastic wrap and allow the dough to rise again in an area where the temperature is 75°–80°F until doubled, about 45 minutes to 1 hour.

Preheat the oven to 475°F about 1 hour before baking. Place the oven shelf at the lowest level.

Turn the dough out onto a cookie sheet and press it down lightly to flatten it slightly. Drizzle the olive oil left in the rising bowl into a loaf pan. Turn the pan to coat all sides with the olive oil. Gently shape the dough into the loaf pan. Cover with plastic wrap and allow to double again, about 45 minutes to 1 hour.

Remove the plastic wrap from the loaf pan and place it in the oven. Bake the bread at 475°F for 10 minutes. Lower the temperature of the oven to 425°F and continue baking for 20 to 30 minutes until the bread is golden brown and sizzling and a skewer inserted into the middle comes out clean.

Remove the bread from the oven and transfer it to a wire rack. Unmold the bread from the pan and allow to cool

for about 10 minutes top side down on the rack. Serve warm while the crust is hot and crisp.

Note: This is my family's favorite winter-day treat, and there is no place better to raise dough than in front of a roaring fire while a snowstorm rages.

Las Delicias Corn Muffins

Something sweet loaded with vegetables and extra fiber? This is another example of what I call a vegetable cake. If you're going to eat sweet, luscious treats—make them healthy! You get the best of both worlds!

MAKES 12

⅔ cup fresh corn kernels cut off the cob
1 tablespoon maple syrup
½ cup stone-ground cornmeal
½ cup flour
3 tablespoons dark brown sugar
1 teaspoon baking powder
¼ teaspoon baking soda
½ teaspoon sea salt
1 large egg
⅔ cup sour cream
2 tablespoons unsalted butter, melted

Preheat the oven to 400°F. In a small bowl, combine the corn kernels with maple syrup. Stir to combine and set aside.

In a separate bowl, whisk together the cornmeal, flour, brown sugar, baking powder, baking soda, and sea salt.

In a separate small bowl, lightly whisk together the egg and sour cream. Stir the egg mixture into the flour mixture until just combined. The batter should be lumpy. Gently fold in the melted butter and corn kernels until just incorporated.

Spoon the batter into muffin cups lined with paper liners, filling them almost to the top. Bake the muffins for 15 to 18 minutes or until the tops are golden brown and a wooden toothpick inserted into the middle comes out clean.

Unmold the muffins and cool them top side up on a wire rack. Serve warm.

Adriance
Farmhouse

Queens County Farm Museum

37-50 Little Neck Pkwy., Floral Park, NY 11004
(718) 347-3276 | www.queensfarm.org

The history of the Queens County Farm Museum, which occupies New York City's largest remaining tract of undisturbed farmland and remains the only working historical farm in the entire city, dates back to 1697. With a sweeping forty-seven acres, it is the longest continuously farmed site in New York State. At its helm: twenty-eight-year-old Kennon Kay.

"Despite our extensive history and carrying the word *museum* in our official name, we're really just trying to be a working urban farm as much as possible," she said. "Although we have forty-seven acres, when you look around you'll see that just about half of our property is wooded." To compensate, Kennon nurtures alpacas, goats, pigs, and chickens that diligently clear the overgrown forest in a perfect permaculture cycle. "New York is huge. People make the mistake of thinking that New York City is just NYC or Manhattan itself. In addition to a lot of neat community gardens and rooftop garden initiatives, there are actually many spaces where people are growing food in the ground on farms."

This particular property was settled by the Dutchman Elbert Adriance when this area was once the lush New Amsterdam. One of the fascinations of the Queens County Farm Museum is its impressive endurance. Farmed continuously through four local wars and two world conflicts, this farm has produced food even as the mega metropolis called New York City grew skyward all around it.

In 1926, New York State purchased the farm, which had been family owned for 250 years, to build the Creedmore State Psychiatric Hospital and used the property for the rehabilitation of patients, for growing food for the patients, and for cultivating ornamental plants and shrubs for the beautification of the campus. Unfortunately, Creedmore demolished all of the

Kennon Kay

original farm buildings except for the farmhouse, destroying all of the valuable historical structures. Today, this Dutch farmstead exists as the Queens County Farm Museum, a New York City landmark on the National Register of Historic Places, providing agritourism and educational opportunities for more than five hundred thousand people every year.

Managing the farm for the last three years, Kennon first traveled around the United States learning farming from her agricultural elders in California and Oregon and Maine. A lovely young lady with sweeping long brown hair, she absolutely dispels the *Saturday Evening Post* stereotype that farmers are ruddy-faced "hicks" or toothless members of the lower classes who wear overalls and chew on straw. "Today there is such a burgeoning movement of young farmers that I feel it is absolutely normal to be a female farmer. I'm on the cutting edge, and I feel incredibly lucky to be at this urban farm. Here I'm just one of a six-woman team." Seven years ago when Kennon started farming, her age and gender were indeed rare in this industry. "My parents thought my choice of agriculture was somewhat weird and were hoping that this was just a rebellious phase," she chuckles. "Thanks to Michael Pollan, they understand why I chose this career and are fully supportive."

In discussions about the value of preserved farmland in urban environments, Kennon knows that this property is truly special. It would have been easy to let this land give way to business and home development. But many people over the years have swum against the tide to fight the pressures of development and to safeguard such an expensive piece of real estate while all of the surrounding land was swallowed by the great paving machines that create livable cities.

To say that the Queens County Farm Museum is unique and valuable is a vast understatement. "It's really special to have a farm of this size in the heart of Queens," Kennon said. "I feel spoiled living in New York City. We have the best food for more than one hundred miles."

Queens Farm Plantain Fritters with Kennon's Mango Sauce

Nothing goes better with plantain fritters than fresh mango sauce loaded with fresh picked herbs right off the farm! Fried plantain is an African dish and served as a snack, a starter, or as a side dish to a main course such as barbecued meat. You can also use fresh peaches or nectarines instead of mangos if you'd like.

MAKES 24 PIECES WITH 1½ CUPS OF SAUCE

Mango Sauce

1 cup peeled, seeded, and cubed fresh mangos
½ cup minced fresh mint leaves
1 jalapeño pepper, seeded and minced
1 small shallot, minced
2 tablespoons brown sugar
½ cup freshly squeezed lime juice
Sea salt and pepper to taste

Fritters

2 large green plantains, peeled
2 cloves garlic, thinly slivered
1 1-inch piece fresh ginger, peeled and thinly slivered
2 tablespoons olive oil
Sea salt and pepper to taste

To make the mango sauce, combine all ingredients in a blender or food processor and puree until smooth. Cover tightly and allow to rest at room temperature for at least 2 hours or overnight in the refrigerator. Refrigerate until ready to use. Shake vigorously before using.

Preheat oven to 400°F. Cover a nonstick baking sheet with foil and coat with nonstick cooking spray.

Grate each plantain on the coarse side of a hand grater. Transfer to a medium bowl and add the garlic and ginger. Toss to mix.

Loosely shape the mixture into 1-inch balls and flatten slightly. Do not pack the plantain shreds too tightly. Arrange the 24 flattened balls on the prepared baking sheet. Brush each with olive oil and season liberally with sea salt and pepper.

Bake for 12 minutes or until golden brown and crispy. Transfer to a platter and serve with Kennon's Mango Sauce.

Two Coves Community Garden

8th Street & Astoria Boulevard, Astoria, NY 11102
(718) 512-8649 | www.twocovescommunitygarden.org

Vanessa Jones-Hall is following in the footsteps of her heroine and mentor Gail Harris as leader of the Two Coves Community Garden on the corner of 8th Avenue and Astoria Boulevard. This bizarrely shaped, four-sided urban garden space, squeezed between weirdly interconnecting roads, used to be a playground for Vanessa and the other children of the Astoria Houses housing project. "We felt like we were away from home," she said, "but in reality, the adults could still look down on us from across the street."

In 2007, Gail started the Two Coves Community Garden to advocate for people who are disabled with physical or emotional challenges and to provide them a creative outlet with flowers or shrubs or trees in this endless steel-frame neighborhood. A mere five years later, this once derelict dump site is an organic community-garden oasis and winner of the Mollie Parnis Dress Up Your Neighborhood Award. "This garden has changed our community in tremendous ways," Vanessa says. "It has not only become a beautiful sight as opposed to the eyesore it once was, but also the collaboration of the ethnicities that work here together every day to keep this garden growing and functional. I'm talking about the different ethnic groups with varied national backgrounds that come to Two Coves Community Garden to share their knowledge of gardening from their own special cultures. Even with the differences of language, we are able to combine our knowledge and love for gardening, and sharing that knowledge brings us all together as one family—a community family. Without this garden, we would have walked past each other on the street without saying 'hello' otherwise. We now come together, harmonize together, and our garden is the result of that collaboration." Rich, poor, Cambodian, Latino—it really doesn't matter. This patch of earth erases the importance of all such defining data.

Vanessa Jones-Hall

Astoria Houses just celebrated its fiftieth year. One of many public housing areas for low- and moderate-income residents in New York City, this development consists of twenty-two buildings, each six or seven stories high, on 32.3 acres. More than 3,500 people live together in 1,102 apartments. Most Section 8 Leased Housing facilities are known popularly as "the projects," and with the concentration of low-income

families, they are also popularly associated with poverty and crime. Astoria Houses, however, is positively associated with farming; community garden space; apple, pear, and fruit trees; and green public space. Fifteen full-fledged members participated in this garden's start four years ago. Today, more than two hundred individual members grow food on their own plots, and an equal number of eager people remain on waiting lists trying to get in. Just think what this neighborhood would look like if every person were provided space to grow their own food.

Two Coves Community Garden is truly a celebrated safe haven. In addition to garden plots, this space functions like a regular brick-and-mortar community center. Harvest festivals and potlucks, Fourth of July barbecues, apple bobbing and pie-eating contests, and tie-dye demonstrations provide amusement, opportunities for dialogue, and human connection in a sunflower-ringed outdoor amphitheater. The Western Queens Compost Initiative recycles residential food waste. With more food being thrown away in NYC than ever before, and with an increasing need for fresh, local, healthy food, this composting program has significant impact on the lives of Astoria's residents, on their health, the soil, their ability to grow food in fertile soil, and the environment by preventing reusable waste from entering landfills.

Vanessa's 100-square-foot plot contains an apple tree for shade and a deck chair for relaxing. All other things are planted in honor of her grandparents, Sally and Willis Jones, and her mother. In 1951, Mr. and Mrs. Jones moved into Astoria Houses when Vanessa's mom was just ten years old. And even as Vanessa grew up in this housing development, she remembers green grass and trees and the beauty surrounding her. While everything has changed now, she keeps the memories of precious loved ones alive in her small, rented garden space.

Seeing an unusual pale green plant in the hands of another gardener, Vanessa stopped for a lesson on kohlrabi, a look of wonder on her face. "Wait. Which part am I supposed to eat?" Having receiving verbal cooking lessons in halting English and with a great deal of encouragement, she walked out of the tall gates to go home for supper with the robust leaves of this mystery vegetable hanging out of her denim sling-shoulder purse.

Vanessa's Apple Crisp

A nod to Vanessa's apple tree in her plot, this apple crisp recipe hails from the 1920s. There is nothing like enjoying the sweet fruits of fall while remembering one's elders. I have a sneaking suspicion that this crisp won't last too long in Vanessa's household. (I know it doesn't in mine.)

SERVES 4-6

Topping

6 tablespoons flour
½ cup packed dark brown sugar
¼ teaspoon cinnamon
¼ teaspoon nutmeg
¼ teaspoon sea salt
5 tablespoons unsalted butter, cut into ½-inch pieces, softened
¾ cup finely chopped walnuts or pecans, toasted

Filling

8 large apples, peeled, cored, and cut into 1-inch cubes
2 tablespoons maple syrup
1½ tablespoons freshly squeezed lemon juice
½ teaspoon grated lemon zest

To make the topping, mix the flour, brown sugar, cinnamon, nutmeg, and sea salt together in a large bowl. Add the butter and toss to coat. Pinch the butter chunks and dry mixture between your fingers until the mixture looks like crumbly wet sand. Add the nuts, toss to incorporate, and refrigerate the topping for at least 15 minutes.

Adjust an oven rack to the lower-middle position and heat the oven to 375°F.

For the filling, mix the apples, maple syrup, lemon juice, and lemon zest together and place in an 8-inch square glass or ceramic baking dish. Distribute the chilled topping evenly over the fruit.

Bake until the fruit is bubbling and the topping is deep golden brown, about 40 minutes. Let cool for 10 minutes before serving.

Note: For a larger amount, double the ingredients and bake in a 9 x 13-inch casserole dish for about 1 hour.

Vanessa with Gail Harris

Two Coves Sun-Dried Pesto Casserole

Who doesn't like macaroni and cheese? Better yet, make this simple dish ahead of time and throw it in the slow cooker. This recipe is a perfect one-pot meal with roasted vegetables or cooked meats tossed in with the tortellini.

SERVES 6–8

2 whole lemons
8 tablespoons unsalted butter, divided
½ cup oil-packed sun-dried tomatoes, drained
2 cups packed basil leaves
2½ cups freshly grated Parmigiano-Reggiano cheese, divided
3 tablespoons all-purpose flour
1½ cups chicken broth
1½ cups heavy cream
1½ pounds fresh spinach-and-cheese tortellini (or other variety)
3 tablespoons black pepper
Zest of 2 lemons, minced
Sea salt and pepper to taste

Coat the insert of a 5- to 7-quart slow cooker with nonstick cooking spray or line it with a slow-cooker liner according to the manufacturer's directions.

Peel 1 lemon and slice crosswise into ¼-inch rings. Sauté lemon slices in 3 tablespoons of butter in a large sauté pan. Remove from the heat and set aside.

Put the tomatoes, basil, 1½ cups Parmigiano-Reggiano, and sautéed lemon slices in the bowl of a food processor. Add the remaining 5 tablespoons of butter and the flour, and pulse until the mixture is completely incorporated.

Transfer the mixture to the large sauté pan used to cook the lemons. Over high heat, add the broth and bring the sauce to a boil, whisking until thickened. Add the cream and bring the sauce back to a boil. Remove from the heat and set aside.

In a large bowl, combine the tortellini, pepper, juice of the second lemon, and the lemon zest.

Spread half the tortellini in the slow cooker. Top with half of the sauce and ½ cup cheese.

Top with layers of the remaining tortellini, sauce, and cheese. Cover and cook on low for 4 to 5 hours, until the pasta is tender and the casserole is bubbling. Remove the cover and cook for an additional 30 minutes.

Serve from the cooker set on warm. Season to taste with sea salt and pepper.

Chapter 7

Rockland County

Camp Hill Farm

126 Camp Hill Rd., Pamona, NY 10970
(845) 362-0207 | www.camphillfarm.org

Small family farms in the suburbs have become an endangered species. Larger, rural farmland properties exceeding thirty-plus acres are the norm, and are aggressively advocated for protection. At the same time, urban farming in dense metropolises is growing at double-digit annual rates. Squashed in between, suburbia suffers the most from loss of tillable terrain. Land and lot sizes have been whittled down to slim margins of acreage surrounding houses. Suburbia does not qualify in size or current-use agriculture infrastructure for the advocacy and protections provided by valuable organizations such as the American Farmland Trust.

In Rockland County, the smallest county in New York State outside of the five NYC boroughs, agriculture was a dominant economic factor into the early twentieth century. By the 1920s Rockland County became a welcome home to artists, composers, actors, and playwrights migrating to this peaceful setting from New York City. Improvements in road systems through the construction of the Palisades Interstate Parkway and the New York State Thruway; a World War II army troop depot, Camp Shanks; and the opening of the Tappan Zee Bridge in 1955 made this county an integral part of the NYC metropolitan area. One of the most negative consequences of this rapid suburban growth, however, was a nearly complete destruction of the county's agricultural history and farming sector.

The US food and farming system contributes nearly $1 trillion to our national economy and employs 17 percent of our work force. In addition to food, well-managed agricultural terrain supplies valuable nonmarket environmental goods and services such as food and cover for wildlife, control and prevention of flooding, and maintenance of clean air quality. Yet it is the farms closest to our cities, but directly in the path of development, that produce 91 percent of our fruit, 78 percent of our vegetables, 67 percent of our dairy, and 54 percent of our poultry and eggs.

Between 1982 and 2007 more than twenty-three million acres of America's agricultural land was consumed by development. Nowhere is this trend more evident than in Rockland County, located twenty-five miles north and west of NYC and north of the seventy boroughs in Bergen County, New Jersey, which is about 8 miles south over the border. In 2000, only five registered farms remained. Today there are nine.

In New York State, land that can be preserved for farming has a typical five-hundred-acre minimum acreage requirement (although many open-space nonprofits are now helping with smaller sites), and the space must be currently farmed to qualify for the purchase of development rights or the implementation of conservation easements. In Rockland County, this restriction has been truly damaging. With the exception of Harriman State Park (in Rockland and Orange Counties),

John McDowell and daughter Luna

there is no longer any open space totaling or exceeding five hundred acres. Additionally, nearly all of Rockland County's land has been zoned residential, commercial, or industrial and is no longer zoned for agriculture.

Farming in suburbia requires our greatest advocacy efforts. To create or preserve new agricultural space in Rockland County, however, is exceedingly difficult. Most farmers in this county are resorting to ingenious uses of their backyards and intensely microfarming on limited-acreage residential plots. There have been no suburban farming advocates. Until now.

Musical composer John McDowell and his wife, Alexandra Spadea, who teaches Rudolf Steiner's Eurythmy, a performance art used in education and as a movement therapy in Waldorf schools, created Rockland's very first CSA program at their backyard Camp Hill Farm.

The Camp Hill Farm is a six-acre, once-forgotten Quaker family farm and farmhouse whose foundations go back to the 1600s. After putting his first hoe in the ground, it didn't take John long to realize that their suburban farming venture needed support and strength in numbers. In response, these artists became the starting momentum behind the Rockland Farm Alliance, a broad-based coalition of farmers, community groups, activists, local and state officials, and everyday citizens. They have come together to preserve, create, and enhance sustainable food production in Rockland County. There is no

handbook on how to save suburban farms. "It's an 'out of sight, out of mind' phenomenon," says Farmer John. "Unless we actively strive toward putting small farms back into suburbia, we'll lose this opportunity forever."

This grassroots effort, which started with the vision of a single backyard farmer, has grown into a powerful community force with over one thousand followers. Rudolf Steiner said, "Eurythmy, from its very nature, is ever seeking for outlet through the human being. Anyone who understands the hand, for example, must be aware that it was not formed merely to lie still and be looked upon. The fingers are quite meaningless when they are inactive. They acquire significance only when they seize things, grasp them, when their passivity is transformed into movement. Their very form reveals the movement inherent within them."

In a way, Camp Hill Farm showcases Steiner's principles. Land, unused and covered with ornamental junegrass, is meaningless when it is inactive. In cultivating the land of Camp Hill Farm, composer John discovered new meaning in this life and added agrarian composition using seedlings to his skilled repertoire of music composition using pencils and staff paper.

Farmer John's Cucumber Salad with Lime Dressing

Most of us roll our eyes at cucumbers. We think of raw slices in salads, brined cucumbers as pickles, or, occasionally, as eye-puffiness remedies if you're really game to lounge around with slices of vegetables on your face. Maybe this will spice things up for you.

SERVES 4-8

1 clove garlic, minced
1 shallot, minced
¾ cup olive oil, divided

⅓ cup freshly squeezed lime juice
½ teaspoon grated lime zest
½ cup tightly packed cilantro stems and leaves
1 jalapeño chile, stemmed, seeded, and minced
1 teaspoon sea salt
½ teaspoon pepper
6 cups thinly sliced cucumbers
Sea salt and pepper to taste

In a small saucepan, sauté garlic and shallot in 2 tablespoons of olive oil until browned and crispy.

In a blender, combine the garlic and shallot, remaining olive oil, lime juice, lime zest, cilantro, jalapeño, sea salt, and pepper in a blender and puree until smooth. Pour into a jar, secure the lid, and refrigerate the dressing until ready to use. Shake well before use.

Toss sliced cucumbers with ¼ cup dressing and allow to marinate in the refrigerator for 1 hour before serving. Season to taste with sea salt and pepper.

Camp Hill Goat Cheese Fettuccine

Goat cheese makes a fabulous base for cheese sauce. Yes, I know . . . traditional cheese sauce recipes call for hard cheeses or cheddar. Try goat cheese! It's creamy and delicious and adds a new flare to traditional pasta and cheese.

SERVES 4

½ pound fresh goat chèvre cheese
⅓ cup grated Parmesan cheese
⅓ cup milk
2 tablespoons minced lime zest
¼ teaspoon sea salt
½ teaspoon freshly ground pepper
1 recipe Phyllis & Eunyoung's Fresh Spun Pasta (see p. 79) or ¾ pound dried fettuccine, uncooked
3 tablespoons freshly squeezed lime juice
Sea salt and pepper to taste
Nancy's Limey Coconut Shrimp (see p. 227)

In a small bowl, combine the goat cheese, Parmesan cheese, milk, lime zest, sea salt, and pepper. Stir to combine.

Meanwhile, in a large pot of boiling, salted water, cook the fresh fettuccine until just done, about 5 minutes, or if using dried pasta, cook according to package instructions. Reserve 1 cup of the pasta-cooking water and drain. Immediately toss the pasta in the fresh lime juice while still draining in the colander.

Whisk the pasta-cooking water into the goat cheese mixture and bring to a rapid boil in the pot used to cook the pasta. Pour the hot goat cheese sauce over the hot pasta in a large bowl and toss to combine. Season to taste with sea salt and pepper and serve immediately on its own or with Nancy's Limey Coconut Shrimp.

Cropsey Community Farm

220 S. Little Tor Rd., New City, NY 10956
(845) 634-4939 | www.rocklandfarm.org

The Rockland Farm Alliance (RFA) is a community coalition that was founded to facilitate local sustainable agriculture and to reinvigorate awareness of the desperate need for these local food resources. Their innovative approach to revive farming in Rockland County has been heralded as cutting edge by state farming authorities. Through the Glynwood Center's Keep Farming process, a program that helps communities explore the feasibility of agriculture, the RFA discovered that opportunities to farm in suburban areas are increasing and that there are many people who both want to farm and who can actually make a living doing so.

The Cropsey Community Farm is the RFA's first farm project, cultivating five acres of the Cropsey Farm. The farm, originally purchased by the Cropsey family in 1893 and actively farmed until 2006, was purchased by Rockland County and the town of Clarkstown through their open-space preservation program. Through a signed partnership with the RFA, the Cropsey Farm became the Cropsey Community Farm and CSA program, thereby canceling the development of forty-eight single-family homes.

In the words of Naomi Camilleri, the executive director of the Rockland Farm Alliance, "In the early 1930s there were nine hundred farms in Rockland County, and we were heavily agrarian, supplying a large amount of produce for New York City. After the opening of the Tappan Zee Bridge in the '50s, it became easier for folks to live in the suburbs and commute to the city. The population of the county exploded, and development became our collective primary focus. Our farms disappeared."

Having grown up in Rockland County, Naomi remembers watching the farmland being bulldozed in the 1970s in favor of strip malls and gas stations. By the year 2000, fewer than five farms remained, and those maintained orchards and didn't produce vegetables. "This

Naomi Camilleri

whole agricultural culture completely disappeared, and people were not even aware of how valuable this farming culture had been. It's a sad situation to see that all of this tradition and history have completely disappeared."

The Cropsey Community Farm stands out like an anomaly in this buzzing, traffic-lit neighborhood. Special homage should be paid and standing ovations given to Jim and Pat Cropsey. Against the growing concrete tides, they continued to farm this land until the late 1990s. And upon retirement, they chose to turn their backs on the daily offers of cold, hard cash from housing developers. Jim and Pat continue living in the property's main house and take pride that their family's farm remains open, green, and fertile under the hands, direction, and management of the Rockland Farm Alliance.

This transition of management hasn't been completely easy or without its challenges. As Naomi moves about the fields with ease and confidence, the Cropseys watch from the window. I asked Jim what this shift of power feels like. "I'm glad not to look out and see housing construction," he said.

The extraordinary speed with which this property transitioned from preserved open space to a working community farm has not always been smooth. This being said, the Cropsey Community Farm is a marvel. Roadblocks will be scaled, problems will be solved, appropriate business systems will be built over the quieter months of winter, and the future of agriculture in New City on this property will be celebrated for many, many years to come. This model, in turn, will spur other agricultural projects throughout Rockland County. And so agriculture begins.

Cropsey Farm Cream of Green Banana Soup

Are you bold enough to eat bananas in a soup? If so, why not make banana soup on February 4, National Homemade Soup Day?

SERVES 8

4 medium-size, very green, hard bananas
4 cups water
1½ teaspoons sea salt, divided
2 cups whole milk
½ cup heavy cream
3 ounces white cheddar cheese, shredded
½ teaspoon white pepper
4 tablespoons freshly squeezed lime juice
Minced fresh cilantro leaves or ground sweet paprika for garnish

Peel the bananas by cutting incisions lengthwise along the ridges and then pulling away the skin. Place bananas in a bowl of cold water to prevent discoloration

In a 4-quart nonreactive saucepan, bring the water and 1 teaspoon of sea salt to a boil. Place the whole bananas into the water and boil, covered, over low heat until the bananas are soft, about 15 minutes. Drain and reserve the cooking liquid. Wash the saucepan.

With a fork, cut each banana into 6 pieces, cutting out any black discoloration in the centers. Place the bananas in a food processor or blender and puree in batches with 2 cups of the reserved cooking liquid until smooth.

Transfer the pureed bananas to the clean saucepan. Add the milk and cream and bring to a boil, stirring constantly. Add extra cooking liquid if the soup appears too thick. Add the cheese and whisk constantly until melted. Remove from the heat and stir in the remaining ½ teaspoon sea salt and the pepper.

Serve in warmed soup bowls garnished with a dash of freshly squeezed lime juice and cilantro or paprika.

Note: When working with bananas, keep in mind that you should not use aluminum utensils because they will discolor the soup. Use green, hard bananas. If the bananas are beginning to yellow, the soup will have a slightly sweet taste, which is also good.

Naomi's Poached Pear Salad

This salad blends the very best of fruits and vegetables and artisanal cheeses! It is a true late-summer celebration all on one plate.

SERVES 4

2 cups apple cider
1 cup dry white wine
1 teaspoon black peppercorns
¼ teaspoon ground cardamom
4 pears, halved lengthwise and cores removed with large spoon
4 slices bacon, cut into ½-inch pieces
4 tablespoons olive oil
1 teaspoon Dijon mustard
2 teaspoons apple cider vinegar
¼ cup thinly sliced red onion
6 dates, pitted and coarsely chopped
8 cups mixed bitter greens: escarole, endive, radicchio, frisee
Sea salt and pepper to taste
½ cup creamy blue cheese, diced
¼ cup toasted pecans, chopped

In a large saucepan bring apple cider, wine, peppercorns, and cardamom to a boil. Reduce heat to a simmer. Place pears, cut-side down, into the poaching liquid. Cover and simmer until tender, about 25 minutes depending on the ripeness of the pears. When cooked, remove from the poaching liquid and let cool to room temperature. Reserve the poaching liquid.

In a medium sauté pan, cook bacon over medium heat until crispy and golden, about 7 minutes. Remove the bacon from the pan and set aside to drain on paper towels. Add ½ cup pear poaching liquid to the hot sauté pan and boil vigorously, stirring to scrape up any browned bacon pieces, until reduced by half. Pour into a large salad bowl. Add olive oil, Dijon mustard, and apple cider vinegar. Whisk to combine. Add onion, dates, mixed greens, and reserved bacon pieces. Toss to combine. Season to taste with sea salt and pepper.

When the pears are cool, stuff the cavity of each with crumbled blue cheese. To serve the salad, divide the greens among 4 plates. Sprinkle with toasted pecans and nestle 2 blue cheese–stuffed pear halves into the greens on each plate.

Hook Mountain Growers

206 Kuyper Dr., Upper Nyack, NY 10960
(917) 880-8264 | www.hookmountaingrowers.com

By the time this book is published, a mini Pamela Yee will have been born and probably will have been taken outside daily to enjoy the chickens in her backyard farm.

Nestled between Hook Mountain and the Hudson River, a mere thirty-minute drive from New York City, lies Pamela's house and homestead—a biodynamic, organic microfarm. Charlie and Pamela, both physicians, originally intended to create nutrient-dense produce to feed themselves and their growing family. Using

Pamela Yee

bio-intensive methods of permaculture, it was quickly apparent that their one-tenth acre of growing space could produce enough food to feed neighbors, shoppers at farmers' markets, and local restaurants. Hook Mountain Growers was born.

In addition to food, Charlie and Pam provide heirloom, unusual, and disease-resistant varieties of vegetables grown using Matt and Teresa Freund's Cow Pots made from the manure by-product of their Connecticut dairy farm. "Nutrient density is the premise that proper care of the soil produces vegetables with maximum nutritional quality," Pam points out. "Take kale, for instance. Kale grown on two different farms with two different growing methods can have two different vitamin and mineral compositions. There are no standard measurements for kale, such as, say, all kale contains 6 percent iron. Nutrient content is directly determined by each kale-grower's soil health."

These busy parents-to-be with demanding day jobs took the time to study under Dan Kitteridge's Real Food Campaign to rebuild the soil underneath their former lawn. This year, in addition to their already blossoming raised beds, they've planted a wild area full of edible plants. Each has a specific purpose to create a balanced soil and terrain ecosystem for the needs of their urban farm. They're growing comfrey, figs, pawpaw, dwarf apple trees and peach trees, mulberry trees, and quince. They've also added a chicken

coop. Instead of pigs or goats or sheep or cows, the Hook Mountain Growers farm's mascot is Henry Hudson the French bulldog.

According to permaculture principles, one plant does not have a singular, stand-alone purpose. Using this, Pamela has created a microfarm, a complete replica of a larger working farm on a miniaturized scale. While most of Pamela's neighbors still mow the grass every Saturday, this family raised 1,600 pounds of food on their one-tenth acre of suburban backyard space. In contrast, most rural farming operations cannot and are not producing 16,000 pounds of food per acre using conventional farming techniques. Amazing.

Pamela, an accomplished cook of Chinese descent, loves growing a plethora of organic Asian specialties not available anywhere else. When she and her husband bought their house, they thought they'd plant herbs, maybe a few heads of lettuce, or perhaps just grow a few vegetables here and there in their spare time. Attending classes at the New York Botanical Garden on seed starting and building raised beds launched a wild farming passion in these two science-loving medical practitioners. They devoured every book and every resource they possibly could. It's fair to say that they were hooked.

For those of you with small one-acre backyards, know that you CAN create a full-scale microfarm. Hook Mountain Growers is an extension of one smart and passionate couple's backyard. Every plant is grown here for a specific reason, and the layout of their space has been modified for maximum soil health and growing conditions. In truth, Pamela's backyard actually looks like something you'd see at the New York Botanical Garden. It is absolutely beautiful. There is a calming serenity upon entering the wooden gates that is, actually, non-farmesque.

"Even one small tomato plant makes a world of difference," Pamela said. "I absolutely love coming home."

Hook Mountain Asian Chicken Noodle Soup

Not all chicken soup is created equal. This Asian-inspired variation is packed with veggies and antibiotics to really help the soul and body feel better. Star anise has been used in tea as a remedy for rheumatism, and the seeds are sometimes chewed after meals to aid digestion. In China, for example, ginger has been used to help digestion and treat stomach upset, diarrhea, and nausea for more than two thousand years. Feel better!

SERVES 6

1 whole chicken, about 4 pounds, trimmed of fat
12 cups stock of your choice (see p. ix) or water
4 shallots, peeled and sliced thin
1-inch piece fresh ginger, peeled, minced, and crushed in a garlic press
4 whole star anise
2 (3-inch) cinnamon sticks
⅓ cup fish sauce or rice wine vinegar
Sea salt and pepper to taste
6 ounces vermicelli, softened in hot water to cover for 20 minutes
2 cups fresh bean sprouts, rinsed and drained
¼ cup minced scallion greens
¼ cup minced fresh cilantro
½ cup minced Thai basil leaves
Lime wedges

Using a heavy knife or a cleaver, cut the chicken through the bones into 10 to 12 pieces.

In a large, heavy stockpot, add the chicken pieces, stock, shallots, ginger, star anise, and cinnamon sticks. Bring to a light simmer over medium heat, skimming the top of the broth, and cook for 1½ to 2 hours or until the chicken is poached through and very tender.

Strain the chicken and seasonings into a large bowl. Return the broth to the stockpot and add fish sauce; cover and simmer gently over low heat. Season to taste with sea salt and pepper.

In a second large saucepan, bring 4 quarts salted water to a boil. Drop the vermicelli noodles and bean sprouts into the water and stir continuously to prevent them from sticking together. Boil for 1 to 2 minutes or until just tender. Drain in a colander under warm running water to remove the starch.

Remove the chicken from the bones and cut or shred the meat into thin strips.

To serve, ladle equal portions of the cooked noodles and bean sprouts into bowls. Top with shredded chicken, scallions, cilantro, and basil leaves. Ladle hot stock into the bowls over the chicken and noodles. Serve with lime wedges.

Pam's Stir-Fried Sweet Potatoes

Stuck with simply baking sweet potatoes? And then what? Load them up with tons of butter like we would a baked russet potato? Here's a new twist on boring sweet potatoes. Too, the cooking time is much faster!

SERVES 6

¾ cup stock of your choice (see p. ix) or water
3½ tablespoons soy sauce
3 tablespoons rice wine or sake
1½ tablespoons rice vinegar
2 teaspoons dark brown sugar
2 teaspoons toasted sesame oil
½ teaspoon pepper
2 teaspoons grapeseed oil
1 orange, zested into fine strips
1 tablespoon peeled and minced fresh ginger
4 sweet potatoes, peeled and cut into ½ x ½ x 3-inch sticks
1½ teaspoons cornstarch mixed with 2 tablespoons cold water
Sea salt and pepper to taste
1 tablespoon minced scallion greens

In a small bowl, whisk together the stock, soy sauce, rice wine, rice vinegar, brown sugar, sesame oil, and pepper. Set aside.

Heat a wok or large skillet, add the grapeseed oil, and heat until very hot. Add orange zest and ginger and sauté for 10 seconds. Add the sweet potatoes and stir-fry for 1 minute, until the potatoes begin to brown without the ginger and orange zest burning. Add stock mixture and bring to a boil. Cover and cook over reduced heat for 10 to 12 minutes or until the potatoes are tender but not too soft. Add the cornstarch slurry and stir continuously. Bring to a boil for 1 minute. Season to taste with sea salt and pepper.

To serve, spoon sweet potatoes onto serving plates and garnish with minced scallions.

Chapter 8

Staten Island (Richmond County)

Decker Farm

435 Richmond Hill Rd., Staten Island, NY 10308
(718) 249-7480 | www.historicrichmondtown.org

David Cavagnaro is the farm manager and executive chef of the eleven-acre Decker Farm, which is part of the Historic Richmond Town, a hundred-acre museum complex on Staten Island. Its main village contains more than thirty original historic structures. Since it is part of the Greenbelt Conservancy, visitors are able to explore the diversity of the American experiences though the lives of ordinary people who struggled, survived, and thrived from the colonial period to the present. The Greenbelt Conservancy mission is to "promote, sustain, and enhance Staten Island's 2,800-acre Greenbelt through education, recreation, conservation and research. The Greenbelt works in partnership with New York City Parks."

David Cavagnaro

Decker Farm is a perfect blend of farm and outdoor restaurant, a partnership that makes sense since Chef David is also Farmer David. His farm stand is nearly always open and features seasonal baked goods, artisan breads, sandwiches, and a plethora of organically grown fruits and vegetables such as figs, heirloom apples, purple cauliflower, assorted mushrooms, garlic, and much more. Chef David also sells his delicious prepared foods. This outdoor, walk-in farm stand showcases perfectly ripe vegetables alongside David's lentil cakes, crab cakes, cream-cheese pound cake, apple strudel, pulled pork, and herb-roasted chicken. You can sip on butternut squash soup or smear fresh goat cheese rolled in herbs on a cranberry-walnut baguette. While a lot of restaurants attempt to bring the farm onto their menus, David and his delightful wife, Pam, bring their culinary menu into the farm.

This 201-year-old farm hosts many monthly special events including an after-school book club, English country dances, pumpkin-picking

Onion
Rolls 1.25 EA.
Semolina
Pumpernickel
Sliders
Brioche
1.25 EA.
Sesame Twist
2-for 1.00
EZ-10
Interfolded
Dry Waxed Paper
Ficcelli 4.00
onion, olive &
walnut
Brickoven white
&
Rustic whole wheat
3.00
STRAWBERRY
Raisin

festivals, and more. On the last Saturday of every month, Decker Farm hosts cooking demonstrations, recipes, and samples. This is a welcome reprieve from the Costco, Trader Joe's, and Staten Island Mall that crowd the edges of these fertile fields.

Not many chefs become farmers. "Being a farmer is new for me. I've had to reinvent myself," David said. "I used to run my own business for fifteen years, but then when the economy failed it was a little tough. At my age, fifty years old, people aren't looking to hire chefs like me and put us back into the kitchen. So, I contacted Ed Wiseman at Decker Farm. I'm able to lease the farm from Historic Richmond Town, and now I am a farmer." After spending long days behind a stove and under grease hoods in restaurant basements, any chef might find the transition to the sun-burning great outdoors and the hours of stooping, bending, and shoveling difficult. Not Dave. He loves this. "Every day is an adventure. Every day is different. I get to work with food and people in a more fulfilling way than I could from behind the stove. Farmers and chefs, we feed people. But, it's different here. I'm different."

He has some fun stories. Dave can make a chateaubriand sauce from memory. Milking his pregnant goat, however, is a whole new skill set. Yes, there are books for that, and this nouveau farmer read them. And yet, some things, like classic French cuisine, are best learned by practice. I asked him how his first goat milking went. Roaring with laughter and with this palm on his forehead he said, "Well, it took three of us, and we spilled the milk." And the goat? "Oh man, she didn't have a good time either."

I can't wait to see how David, Pam, and Decker Farm continue to grow. Their future is bright and exciting indeed.

Decker Farm Pale Ale Rabbit Stew

While this recipe calls for rabbit, which I'd recommend that you try sometime, you can use any other meat that you'd like! Try this with chicken or ribs or anything else that strikes your fancy!

SERVES 2-4

1 whole rabbit, about 5-7 pounds
1 teaspoon sea salt
1 teaspoon freshly ground black pepper
4 tablespoons olive oil
1 large onion, sliced thin
10 juniper berries
2 tablespoons whole white peppercorns
1 tablespoon whole mustard seeds
2 cups peeled, diced sweet potatoes
1 cup halved dried prunes
4 large sprigs thyme, or 6–8 small ones
1 grapefruit, peeled and cut into wedges
1½ (11.2-ounce) bottles Pale Ale, divided
1 tablespoon cornstarch
2 tablespoons cold water
Sea salt and pepper to taste
1 tablespoon grapefruit zest
2 tablespoons chopped fresh parsley

Preheat the oven to 345°F. Liberally rub the rabbit with sea salt and pepper. In a large sauté pan, heat the olive oil and sauté the rabbit over medium heat for about 5 minutes on each side, or until it is nicely browned. Remove the rabbit from the pan and set aside.

To the hot sauté pan, add the onions and sauté until caramelized and browned, about 8 minutes.

Combine the juniper berries and the whole peppercorns in a small sachet or a piece of cheesecloth tied with a string.

In the bottom of a large oval baking dish with a lid, layer the mustard seeds, sweet potatoes, prunes, thyme, and grapefruit. Nestle the rabbit on top. Pour the Pale Ale over the rabbit and tuck the sachet containing the juniper berries and peppercorns down into the liquid.

Cover the dish, and bake in the oven at 345°F for 30 minutes. Reduce the heat to 320°F and continue baking for 30 minutes or until tender. Remove the lid from the casserole dish, reduce the heat to 300°F, and continue baking for about 20 minutes more.

Remove the rabbit from the roasting pan. Cut the meat into 4 pieces and place on a large platter. Strain the juices from the roasting pan and place half of the cooked sweet potato mixture into a blender, removing the sachet and thyme sprigs. Arrange the remaining sweet potato mixture on the platter around the rabbit.

In the blender, puree the cooking juices and grapefruit mixture until smooth. Strain through a sieve into a small sauté pan. Bring the sauce to a boil and thicken with cornstarch mixed with 2 tablespoons cold water. Boil for about 30 seconds and season to taste with sea salt and pepper. Immediately add the grapefruit zest and pour over the platter. Sprinkle with parsley and serve.

Farmer Dave's Mango-Roasted Pork Shoulder

Feel free to cook with dried fruit—in fact, we don't do it often enough. Dried fruit is a festive way to add new flavor combinations, vitamins, minerals, and fiber to our old comfort food stand-by favorites. Don't like mangos? Try apricots or pineapple or prunes or cranberries or whatever you feel like. Experiment!

SERVES 6-8

One 6- to 7-pound fresh, bone-in picnic shoulder roast or Boston butt, with the skin
1 whole garlic head, separated and peeled
½ teaspoon paprika
1 teaspoon yellow mustard seeds
2 tablespoons coarse sea salt
1 large onion, thinly sliced
1 medium fennel bulb, trimmed and thinly sliced
1 cup dried mango strips
2 tablespoons green cardamom seeds
1 tablespoon whole fennel seeds
3 tablespoons black peppercorns
1 cup fresh thyme sprigs, about one handful
4 tablespoons hot red pepper flakes
2 cups orange juice
2 cups chicken stock
1 tablespoon cold unsalted butter
¼ cup Cognac
Sea salt and pepper to taste

Preheat the oven to 450°F. Wash and dry the pork roast. Score the skin of the pork in a crisscross diamond pattern (like a ham), making deep cuts about 1-inch apart.

Using a mortar and pestle, mash the garlic, paprika, mustard seeds, and coarse sea salt to make a thick paste. Rub all over the pork roast. Set aside to warm slightly at room temperature.

In a large roasting pan, sprinkle onion, fennel, mango, cardamom, fennel seeds, black peppercorns, thyme, and hot red pepper flakes to cover the bottom of the pan.

Place the pork roast, skin side up, directly on the onion, fennel, and spice mixture in the roasting pan.

Combine the orange juice and chicken stock. Add about one-third of the mixture to the bottom of the roasting pan, enough to cover the onion, fennel, and spice mixture.

Roast the pork for about 45 minutes at 450°F until deep golden brown. Add the remaining orange juice and stock mixture and baste the roast thoroughly.

Reduce the temperature of the oven to 250°F and, basting the roast often, cook until a probe thermometer inserted in the center of the meat reads 150°F, about 4 to 6 hours.

Remove the meat from the oven and move the roast to a cutting board. Tent loosely with foil and allow to rest for about 20 minutes.

Transfer the cooking liquid, onions, and spices, to a blender. Skim off as much fat as possible. Puree the mixture until smooth, and while the blender is running, add cold butter and Cognac.

Slice the meat across the grain and arrange on a platter. Pour the sauce over the meat and serve warm. Season to taste with sea salt and pepper.

Greenbelt Conservancy

200 Nevada Ave., Staten Island, NY 10306
(718) 667-2165 | www.sigreenbelt.org

The Greenbelt Conservancy is 2,800 acres of mature forests, wetlands, parks, and meadows in the heart of Staten Island. It includes New York City's largest remaining forest preserve, offering an oasis of peace, quiet, solitude, and wild foods. The Conservancy is also home to tidal and freshwater wetlands, oak and beech forests, and thirty-five miles of hiking trails. Kathleen Vorwick is the president and senior manager of the Greenbelt Conservancy, which leads the entire nation in open-space preservation in urbanized areas. A square city park can only provide a green view to the houses directly neighboring it. In contrast, because the Greenbelt Conservancy snakes throughout the heart of Staten Island, thousands of homes and businesses and schools reap the benefits of green parkland exposure along its infinite, twisting borders.

If you visit the Greenbelt Conservancy, you can become your own hunter-gatherer farmer. Hunting for wild mushrooms involves a lot of common sense. The mushroom season in New York City extends from April through November. According to the New York Mycological Society, there are well over one hundred varieties of edible mushrooms in High Rock Park alone, including those breathtakingly tasty and elusive, golden morels. The old saying goes, "Do not ask me where to find morel mushrooms, and I will not lie." In truth, most mushroom hunters have their own personalized "SWATS"—and yes, this is an abbreviation for a technical term: Scientific Wild Ass Theories—on how, when, and where to find the RIGHT mushrooms. Having said that, it is always wise to connect with a mushroom-foraging group, study the literature, carry a camera, and locate an expert to help you identify the mushrooms that you have picked as a second fail-safe precaution before you eat them.

The Greenbelt Conservancy offers many more food-stalking dining options, such as fiddleheads, ramps and wild leeks, pawpaws,

Kathleen Vorwick

common sow thistle, American hackberry, beach plums, black walnuts, cuckoo flowers, mulberries, and much, much more. "Wildman" Steve Brill is one of NYC's go-to guys for foraging. His tour calendar offers field walks throughout the Northeast. He's famous for being arrested by undercover NYC park rangers for eating a dandelion in Central Park (the Associated Press headline read "Planted Decoys Nab Foraging Botanist") and for a follow-up discussion in 1986 about foraging for food in city parks with Dan Rather on the *CBS Evening News*. Today, smart and educated foraging throughout the NYC parks system is common practice. If you're trying something for the first time, taste a tiny bit of it and then wait twenty-four hours. Know that many plants are Jekyll and Hydes: Depending on the time of year, a plant may be edible one month and then poisonous later in the year.

Thanks to the Greenbelt Native Plant Center, these edible native plants are preserved, repopulated, and protected. What better way to use the Greenbelt Conservancy's many trails than for hikes, learning about and foraging for edible plants with the experts, and then picnicking with your newly found lunch?

The Greenbelt Conservancy also offers the famous Carousel for All Children and other family options such as summer camps, family craft workshops, short family hikes, and kid-friendly musical performances. In addition, a variety of year-round events are available, such as guided hikes, book talks, and art workshops for adults to socialize, network, explore, and celebrate this preserved open space in the heart of Staten Island. A women's walking-and-hiking social club offers opportunities to get in shape, redefine exercise routines, and enjoy the great outdoors with new

friends. A primitive skills and survival group teaches empowerment, awareness, and respect for living off the land, survival skills, and self-reliance in the woods. And, if you have a large dog, there are doggie play dates and clubs especially for you and your dog to enjoy. Or, if you'd rather play golf, baseball, tennis, or soccer or go birding or visit the archery fields, well, then, the Greenbelt Conservancy has land for that. You can visit the Museum of Tibetan Art, the Model Airplane Field, Moses' Mountain, a recreation center, the William T. Davis Wildlife Refuge, and the famous Historic Richmond Town. Kathleen is waiting for you.

Most important, for all that the Greenbelt Conservancy currently is, more plots and parcels of land on Staten Island are actively being advocated for preservation. While enjoying any number of the many wide-open habitats, activities, and food-foraging opportunities, know that the Conservancy is an active project that still needs support, your awareness, and financial donations to purchase additional lands into the Greenbelt family. As our elders advocated for this space that we know today, so, too, should we assist the Greenbelt Conservancy to grow larger and stronger for the generations ahead.

Kathleen's Wild Watercress Salad

Looking for something that takes less than 20 minutes and can stay "on hold" while company arrives or that can sit unattended on a large buffet table and still look beautiful? Look no further.

SERVES 4

1½ tablespoons freshly squeezed lime juice
3 tablespoons olive oil
½ teaspoon sea salt
½ teaspoon pepper
½ cup finely shredded watercress, thick stems trimmed
½ cup finely sliced sweet onion
1 cup peeled and finely grated carrot
2 cups finely shredded green cabbage
12 cherry tomatoes, halved
1 tablespoon finely chopped cilantro leaves
Sea salt and pepper to taste
3 tablespoons crushed, toasted walnuts

To make the salad dressing, whisk together the lime juice, olive oil, sea salt, and pepper.

In a salad bowl, toss together watercress, onion, carrot, cabbage, tomatoes, and cilantro. Season to taste with sea salt and pepper. Fold in the dressing and serve sprinkled with walnuts.

Greenbelt Seared Mushroom Salad

SERVES 4

2 tablespoons unsalted butter
3 cups thinly sliced mixed mushrooms
1 tablespoon fresh thyme
½ teaspoon sea salt
1 tablespoon soy sauce
¼ cup dry white wine
Sea salt and pepper to taste
1 teaspoon Dijon mustard
2 teaspoons freshly squeezed grapefruit juice
3 tablespoons olive oil
2 cups torn arugula (bite-size pieces)
2 cups torn escarole (bite-size pieces)
¼ cup finely crumbled aged white cheddar
2 tablespoons coarsely chopped walnuts

In a large sauté pan, melt butter over medium-high heat. Add mushrooms, thyme, and sea salt. Sauté for 5 minutes, or until the mushrooms begin to soften and release their water.

Add soy sauce and white wine and continue sautéing for 5 minutes, or until the mushrooms are golden. Season to taste with sea salt and pepper.

In a medium bowl whisk together mustard, grapefruit juice, and olive oil to make a vinaigrette. Add greens and toss.

Divide the greens onto plates and top each with sautéed mushrooms and sprinkle with cheddar and walnuts. Season to taste with sea salt and pepper.

Greenbelt Native Plant Center

3808 Victory Blvd., Staten Island, NY 10314
(718) 370-9044 | www.greenbeltnativeplantcenter.org

The Greenbelt Native Plant Center (GNPC) is a facility of the New York City Department of Parks & Recreation. It is a thirteen-acre greenhouse, conservation nursery, seed-saving, and seed-bank storage complex providing locally appropriate seeds and plants to appropriate habitats throughout New York City. Its primary goal is to propagate, plant, and replant indigenous vegetation to restore and help manage the city's valuable natural areas.

There are more than two thousand different species of native plants in the greater New York City area. Ed Toth, director of the Greenbelt Native Plant Center, is a grass farmer—or a lizard's tail farmer, or a swamp milkweed farmer, or a wild bergamot farmer. It depends really, and it changes nearly every day.

Restoration of degraded land requires a science-based approach, and using plant materials from local populations ensures the success of reemerging natural ecosystems. The Greenbelt Native Plant Center's seed collection process begins with research for the right plants for seed harvest, conservative collection of those seeds so as to maintain existing wild populations, planting of these seeds in select locations on their compound, harvesting of new seeds from these grown plants, saving them, and then dispensing these seeds throughout the NYC metropolis.

In the early 1980s, a trio of visionary Staten Islanders, together with naturalists, botanists, and ecological educators, created the Greenbelt Native Plant Center out of a growing concern for the rapidly diminishing natural areas of Staten Island. Their first mission involved locating and rescuing endangered native plants at risk of being bulldozed by commercial development. Today, Ed Toth is farming close to fifty native and endangered species in "pure stands"—the seeds of which are sent back out into New York City to prevent them from becoming urban-endangered species in the future.

The Greenbelt Native Plant Center is a farm, albeit of a very different sort than we're used to. The property on which their facilities were built also has extensive agrarian roots as the former Mohlenhoff family farm. Carl Mohlenhoff, a

Ed Toth

retired NYC Parks employee and nursery manager of the Greenbelt Native Plant Center, began and ended his career on the same land where he was born and raised. Carl's grandparents, Henry and Wilhemine, had twelve robust children, and the birth of each one brought additions to the ever-expanding family home, creating a heftily outsized building that now hosts the GNPC offices and its employees. As the children grew and married, adjoining houses were built around the farm's property. The Mohlenhoffs are remembered fondly across New York City. The entire family, all three generations, belonged to the same church and spent every holiday together at the main house. There was respect and harmony throughout this family, both in their relationships and in their business dealings. Although this property, now under the management of Ed Toth, grows owl fruit sedge and sheep laurel and wild yam (instead of carrots, celery, or lettuce), the Mohlenhoff family legacy, business ethos, and sense of community giving are an integral, daily part of the Greenbelt Native Plant Center.

"What we do here is exceedingly important," Ed says, "because New York City is especially urbanized. We still have functioning natural areas in the city, and we still have a tremendous diversity of species and over twenty-five different plant habitats with thousands of different plants that can still survive here in our urban centers.

At one point, we had three to four times the number of different species than we're left with now. The city has built out, habitats have been destroyed, and all we're left with now are public lands. These lands are becoming increasingly more valuable for both their biodiversity values and also their incredible importance to the well-being of New York City. These native plants in our remaining open public areas provide water filtration and clean air and help lower the temperature of the city. There is nature in urban areas. There is a function to that nature that many people didn't understand twenty-five years ago. Now, before we lose the biodiversity and sustainability of our natural areas, we've launched into conservation mode. Maintaining native plants in urban areas is critical to the health of NYC and its residents."

Mohlenhoff's Turkey Stock

While labeled as turkey stock, this recipe works with all poultry bones. Chicken or turkey stock has never been easier—throw it in the slow cooker and go!

MAKES 8 CUPS

Bones from one cooked turkey
2 medium onions, coarsely chopped
3 medium carrots, coarsely chopped
3 sweet potatoes, peeled and coarsely chopped
8 cups chicken stock
1 small handful fresh thyme (stems and leaves)
2 bay leaves
6 whole cloves
10 black peppercorns
Sea salt and pepper to taste

Put all of the ingredients except the sea salt and pepper in a 7-quart slow cooker. Cover and cook on high for 4 to 5 hours or on low for 8 to 10 hours.

Strain the stock through a colander to remove the large solids. Strain again through a fine-mesh sieve.

Season to taste with sea salt and pepper. Cool the stock to room temperature, remove fat that forms on the top after cooling, and then store in airtight containers in the refrigerator for up to 5 days or in the freezer for up to 6 months.

Ed's Honey-Grilled Salmon with Almond Vinaigrette

This vinaigrette works wonders for just about anything seared or grilled, so keep the recipe handy even if you're not particularly in the mood for salmon.

SERVES 4

Vinaigrette

⅓ cup almonds, toasted
4 teaspoons freshly squeezed orange juice
1 shallot, peeled and minced
2 teaspoons honey
1 teaspoon Dijon mustard
⅓ cup olive oil
1 tablespoon heavy cream or plain yogurt
1 tablespoon chopped fresh parsley leaves
Sea salt and pepper to taste

Salmon

4 (4-ounce) skin-on salmon fillets
1 tablespoon vegetable oil
1 tablespoon olive oil
1 tablespoon honey
½ teaspoon sea salt
½ teaspoon pepper
1 lemon, cut into wedges for serving

To make the vinaigrette, place toasted almonds in a ziplock bag and, using a rolling pin, pound until no pieces larger than ½ inch remain. In a bowl, combine almonds, orange juice, shallot, honey, and mustard. Whisk in the olive oil to create an emulsion. Add the cream and parsley. Stir to combine. Season to taste with sea salt and pepper.

Place the salmon fillets skin side up on a rimmed baking sheet or large plate lined with paper towels. Place more paper towels on top and press down to extract as much liquid as possible. Wrap the salmon in clean paper towels and let dry for at least 20 minutes at room temperature.

When the grill is hot and ready for cooking, lightly grease the grill grates with vegetable oil. In a small bowl, combine the olive oil and honey and brush over the salmon fillets, avoiding the skin. Season with sea salt and pepper. Place the salmon on the grill, skin side down, at a diagonal angle to the grill's grates. Cover and cook without moving until the skin side is brown, well marked, and crisp, about 3 to 5 minutes. If the salmon does not lift off of the grill, continue cooking and checking at 30-second intervals until the fillets release. Using two spatulas, turn the fillets over and cook, covered, until the centers of the fillets are opaque and register 125°F, about 2 to 6 minutes longer.

Serve immediately with lemon wedges and drizzled with vinaigrette.

Joe Holzka Community Garden

1171-75 Castleton Ave. & Barker St., Staten Island, NY 10310
(718) 816-9331 | www.greenthumbnyc.org

As Katie Terlonge puts it, "I've done met my new favorite friend and have fallen in love." She's talking about Miss Spotty, the white-and-tan rabbit and the newest member of the Joe Holzka gardening family. While checking on the green peppers, she hums sweet nothings of friendship and encouragement to Miss Spotty, who is wrapped tightly in the front of her T-shirt. It's easy to see that this community garden is beyond special.

Attorney Joseph Johnson Holzka (1927–1993), a lifelong resident of Staten Island, was a prominent figure in social and political spheres throughout New York City and is most celebrated posthumously as a founding member and avid supporter of the Greenbelt Conservancy (see page 185). In his honor, the Joe Holzka Community Garden was created on the old fire-charred vacant lot where the Plaza Casino once stood. After the fire and the extensive years in which the space existed as nothing more than an eye-sore

garbage dump, the Neighborhood Housing Service (NHS) leased the site to create a community garden that would provide aid to homeowners and revitalize this declining neighborhood.

Managed by Cathy Zelonis and Katie Terlonge, the Joe Holzka Community Garden hosts twenty-three intergenerational and multinational food-growing families on less than a single acre. Community members, future garden members, and city-servicing nonprofits all collaborated to rehabilitate this dilapidated space. Plots surrounded by chrysanthemums, marigolds, and impatiens can be leased free of charge by local residents. The blazingly white gazebo offers shade, a place for tea, or an opportunity to rest and chat about the weather, or politics, or neighbors, or literature.

While I was touring the grounds, Lenny Librizzi, the assistant director of Open Space Greening with GrowNYC, peddled his uber-funky city bicycle into the garden to check on their state-of-the-art rainwater-collection system. This knot of tanks and pipes and valves prevents a tremendous amount of rainwater from washing away into the sewers and out into the ocean and collects it for usage by all of the Joe Holzka Community Garden members. "GrowNYC installed these rain collection systems in twenty gardens throughout all five boroughs," Lenny said. "Members don't have to retrieve water from the hydrants. These are fantastic systems."

Katie Terlonge

No natural resource is more important to community gardeners than simple water. After the distressing droughts in 2001 that restricted gardeners' access to hydrants, GrowNYC and GreenThumb founded the Water Resources Group and a rainwater harvesting team. To date, they have built eighty rainwater harvesting systems that collect over a million gallons of rainwater annually. The Joe Holzka Community Garden is the only lucky recipient on Staten Island.

GrowNYC is a hands-on nonprofit that improves New York City's quality of life through environmental programs and open-space projects to ensure a clean and healthy environment for future generations. They provide access to healthy, fresh, and local food, grow and maintain vibrant green spaces and community gardens, help New Yorkers recycle and reduce waste, and cultivate the next generation of environmental leaders through hands-on educational programming.

"If you walk into a community garden, you'll notice that the temperature is actually a few degrees cooler than it is just a few feet away out on the street," Lenny points out. Other statistics demonstrating benefits to the environment underscore the importance of the Joe

Holzka Community Garden, such as acting as a community composting site to divert garbage from landfills. But Katie sums up the human component: "I just love this place. It's good for my health," she laughs. "I just absolutely love this place." She grins as she sweeps her arms to show me all the parts of the garden that are "hers"—all the parts that belong to her, that are under her control, and that she raises up from nothing. At seventy-three years of age, Katie appears to fit into the typical socioeconomic and cultural demographic in this impoverished neighborhood. Yet, Katie stands with her back straight, her head high as she joyfully marches through her neighborhood handing out vegetables. "I raise it and give it away," Katie smiles. Lenny smiles, too, and looks at me with satisfaction. How can you not feel satisfaction when Katie starts to walk down the street on her way home holding Miss Spotty in her shirt front and handing out green peppers to everyone she passes?

Joe Holzka Green Curry Paste

A good curry paste is hard to find, and few people make their own. A good curry paste is a much-needed skill for any good cook and it takes many times to master. Extra paste can be frozen for up to 2 months. To taste the paste during preparation and to adjust seasonings to your liking, mix 1 teaspoon of paste with ¼ cup of milk.

ENOUGH FOR 4 CURRY DINNERS

10 medium cloves garlic, peeled and put through a garlic press
1 teaspoon sea salt
½ cup minced cilantro stems
2 tablespoons ground coriander
2 teaspoons ground cumin
½ teaspoon ground white pepper
¼ cup stemmed, seeded, and coarsely chopped Thai chiles or green serranos
¾ cup stemmed, seeded, and coarsely chopped green jalapeño chiles
3 tablespoons peeled and minced fresh ginger
½ cup minced fresh lemongrass
4 teaspoons minced lime zest
4 teaspoons minced orange zest
6 tablespoons minced shallots
½ teaspoon shrimp paste or anchovy paste (optional)
1 tablespoon maple syrup
2 tablespoons coconut cream or olive oil
Sea salt and pepper to taste

Place all ingredients (except sea salt and pepper) in a food processor and puree. Stop occasionally to push down the ingredients, and continue pureeing until a thick paste has formed. Season to taste with sea salt and pepper.

Store in a covered glass container or bowl in the refrigerator for up to 1 month. Use with your favorite recipes or as an aromatic condiment.

Cathy & Katie's Thai Chicken Curry

I often wake in the night dreaming of hot curry reminiscent of my days in Seattle. This recipe yields a medium-hot curry. Add or lessen the green curry paste according to your tastes. If using a premade commercial green curry paste, use 2 tablespoons. Also, if you're not particularly in the mood for chicken, use whatever you'd like. If using shrimp, add it toward the end to prevent overcooking. For an all-vegetable curry, omit the chicken and increase the amount of vegetables to 8 total cups of any type or combination.

SERVES 4

2 (14-ounce) cans coconut milk, not shaken
½ cup Joe Holzka Green Curry Paste (at left)
2 tablespoons fish sauce
3 tablespoons maple syrup
1½ pounds chicken breasts, cut into ½-inch cubes
½ teaspoon sea salt
½ teaspoon pepper
3 cups peeled and diced Japanese eggplants
1 cup pineapple chunks
1 large handful snow peas, trimmed
1 tablespoon freshly squeezed lime juice
½ cup whole fresh basil leaves
½ cup whole fresh mint leaves
Sea salt and pepper to taste
Cooked jasmine rice or rice noodles

Carefully spoon off about 1 cup of the top layer of coconut cream from a can of coconut milk; this will be thick and possibly solid. Place the coconut cream and curry paste in a large pot and bring to a boil over high heat, whisking to blend continuously until all of the liquid evaporates and the cream separates into a puddle of colored oil and coconut solids, about 10 minutes.

Whisk in the remaining coconut milk, fish sauce, and maple syrup, and bring back to a simmer. Cook, stirring continuously until the sauce thickens, about 5 minutes.

Season the chicken with sea salt and pepper and stir it into the pot until the pieces are separated and evenly coated with the sauce. Add the eggplant and simmer over medium heat for 5 to 7 minutes or until the eggplant starts to get tender. Add the pineapple and snow peas and continue to simmer until the chicken is cooked, about another 7 to 8 minutes.

Move the pot off the heat, stir in lime juice, basil, and mint, and season to taste with sea salt and pepper. Serve immediately with jasmine rice or rice noodles.

Chapter 9

Westchester County

Blum Center for Health

34 Rye Ridge Plaza, Rye Brook, NY 10573
(914) 652-7800 | www.blumcenterforhealth.com

No two people have the same genetic makeup, history, or life experiences. Each person is unique and different, with individualized pasts, toxin exposures, food histories, emotional reactions, lifestyles, and stress-coping mechanisms. It makes sense that our medical care should be as specialized as our own exclusive, determinate needs. An individually tailored health plan that addresses one's specific personal history is called functional medicine, and its overall goal is not only to improve your health, but also to prevent future medical issues. The Blum Center for Health is a patient-centered medical clinic that addresses the whole person and not just randomly isolated sets of symptoms.

The Blum Center for Health was created by Dr. Susan Blum, an assistant clinical professor in the Department of Preventive Medicine at the Mount Sinai School of Medicine. Her passion lies in identifying and addressing the root causes of chronic illnesses. Her trusty sidekick is Chef Marti Wolfson, who teaches cooking classes at the center's kitchen, Nutrition@BlumKitchen.

A holistic approach to medicine sometimes carries with it the stereotype of "sackcloth and ashes." A chef and nutrition educator teaching cooking classes could give some people the impression that the menu will consist of raw carrots—raw carrots and that's it! Yum! Hope you're not still hungry! Put those notions aside. I assure you that both the Blum Center for Health and its resident chef will dispel every such negative stereotype that may prevail.

We've all heard that if you'd like to lose weight, you should eat more vegetables. This argument is usually based on the fact that vegetables simply have fewer calories than processed foods. But the medical approach that uses "food as medicine" is much more than this generic oversimplification. Fruits and vegetables contain a plethora of vitamins, minerals, and enzymes that our bodies directly need to function properly, and without these, the body is more prone to weight gain and illnesses.

Marti Wolfson is a grinning, confident young lady with a soft voice and mountains of perfectly curly hair. One is instantly at ease in her presence. As a chef, she builds seasonal menus and teaches cooking classes on just about every topic vegetarian or nonvegetarian: Mediterranean cuisine, healthy holiday snacks, ancient grains, or sweet apple desserts other than apple pie. She caters farm-to-table dinners, special events, and corporate cooking and wellness classes using local food.

The Blum Center for Health takes food as medicine one step further. Chef Marti works directly with local farms and Mike's Organic Delivery from Greenwich, Connecticut. "Where we're getting our food is crucially important," Chef Marti said. "It has been shown that locally grown food is more nutrient dense. The farther that food has to travel, the more vitamins,

blum MD CENTER FOR HEALTH
34
Rye Ridge Plaza

minerals, and enzymes that are lost in the pre-picking before that product has correctly ripened, in the cold-storage shipping and transportation, and through the re-ripening process at its final destination." Mike Geller of Mike's Organic Delivery agrees. "Knowing where your food comes from and how it was raised are the most important parts of eating healthily." To that end, Mike's Organic Delivery has close relationships with the best, brightest, and most reputable small organic farms throughout the Northeast. Every day, he drives out to their farms and inspects their produce, veggies, eggs, meats, honeys, syrups, and value-added products to ensure the highest quality. And then, he delivers all such things to his multitude of clients, including Chef Marti Wolfson.

Incidentally, we inherently determine nutritional quality by taste, aroma, and flavor. We prefer the sweet melt-in-your-mouth delight of field-ripened strawberries. We ooh and aah as juice runs down our chins while bemoaning the cardboard-tasting strawberries sold in our grocery stores in December. The difference in taste between the two fruits is determined by the quantity and quality of its vitamins, minerals, and enzymes. The Blum Center for Health should be nicknamed the "Dr. Local Food Clinic."

Blum Center's Gingered Cranberry Sauce

Oftentimes we eat food, simply because it is food and we don't always understand or know the health benefits of what we're eating. Chef Marti can tell you. In the meantime, know that this recipe is power-packed with cold- and flu-fighting antioxidants!

MAKES 3 CUPS

2 cups raw, whole cranberries
2 medium oranges, peeled and seeded
¼ cup honey
2 teaspoons lemon juice
1 cup halved seedless purple grapes
1 teaspoon of freshly minced ginger
Sea salt and pepper to taste
2 tablespoons fresh thyme leaves

Combine cranberries, oranges, honey, lemon juice, grapes, and minced ginger in a food processor and pulse until uniformly chopped and smooth. Season to taste with sea salt and pepper.

Stir in fresh thyme leaves and serve as a condiment to your favorite meal.

Marti's Baked Squash with Mint Sauce

This dish hails from Syrian Jewish roots. There is always that joke about locking your windows and doors during zucchini season. I'd recommend that you put up a big sign asking for more and enjoy the flavors of the Middle East while you're at it.

SERVES 6

1 (15.5-ounce) can chickpeas, drained and rinsed, or ¾ cup dried chickpeas
1 cup long-grain white rice, soaked for 30 minutes in cold water to cover, drained, and rinsed
1 cup diced onions
½ cup melted unsalted butter, divided
2 teaspoons sea salt, divided
2 teaspoons freshly ground black pepper, divided
¼ teaspoon cinnamon
½ teaspoon ground allspice
3 medium-size yellow summer squash, 7–8 inches long
3 medium-size zucchini, 7–8 inches long
⅔ cup fresh lemon juice
2 teaspoons fresh chopped garlic
¼ cup minced fresh mint leaves
Sea salt and pepper to taste

If using canned chickpeas, proceed to the next step. If using dried chickpeas, soak them in water to cover overnight. Drain and rinse the beans, and place them in a large pot with enough cold water to cover by 2 inches. Bring to a boil over high heat. Reduce heat to medium, and cook at a slow boil until fork tender, about 1¼ hours. Drain and transfer to a large bowl.

Preheat oven to 350°F.

In another large bowl, mix the drained chickpeas, rice, onions, 3 tablespoons melted butter, 1 teaspoon sea salt, 1 teaspoon pepper, cinnamon, and allspice. Stir to combine.

To prepare the squash and zucchini, cut each in half crosswise leaving the ends intact. Do not cut the squash lengthwise to create a boat shape. Using a vegetable corer, scoop out all of the seeds, leaving ¼ inch of flesh lining the squash and zucchini shells.

Marti Wolfson

Stuff each squash with chickpea mixture to within 1 inch of the end; the filling will expand during cooking. Press the filling firmly into each squash.

Grease the inside of a large casserole dish with a tight-fitting lid with 1 tablespoon butter. Place the stuffed squash tightly together, side-by-side, in a single layer. Cover the casserole with the lid and bake in the oven for 30 minutes.

In a small bowl, combine lemon juice, garlic, the remaining melted butter, the remaining sea salt and pepper, and the fresh mint. Remove the lid from the casserole dish and pour the sauce over the stuffed squash. Continue to cook, uncovered, for another 30 to 45 minutes or until the rice is tender and the sauce is browned and bubbling. Season to taste with sea salt and pepper. Serve warm or chilled and sliced into rings.

LANE
ONLY
BUS
6720
BUS
LANE
FIRE
ONLY
BUS

Farm Share

70 Hillcrest Ave., Larchmont, NY 10538
(914) 315-1851 | www.myfarmshare.com

Gail Brussel connects families to local agriculture by bringing the farmer directly to their doorsteps. She assists more than four hundred busy people who simply can't always get to their community supported agriculture (CSA) pick-up points to retrieve their weekly shares of food. This service helps active farms reach more customers who typically can't purchase CSA shares because of time and commuting-nightmare travel constraints. In short, Gail is a petite and savvy middlewoman.

Officially, Farm Share is a CSA delivery service that brings freshly harvested, locally grown crops directly to your doorstep. With Farm Share, you become a "member" of Norwich Meadows Farm during the CSA season that you choose (either summer or winter or both), and you'll receive a delivery of your portion of that farm's weekly harvest. As each season changes, so, too, will the diversity of your weekly produce. Founded in 2006, this revolutionary delivery service remains a family business operated by Gail Brussel. In addition to weekly deliveries, Gail also provides recipes created by member chefs and an online recipe exchange on her website for continued learning.

Gail sources her CSA products from Norwich Meadows Farm, which started in 1998 on a half-acre parcel behind Zaid and Haifa Kurdieh's house in Norwich, New York. This small-scale organic farm has grown exponentially to greater than fifty acres. Zaid and Haifa now deliver their products through a multitude of business partnerships, like Farm Share, reaching outside of their Chenango County geographic area. Norwich Meadows Farm is a story of land and loss, diaspora, and of younger generations reclaiming their ancestral farming history through the cultivation of new lands. Both Zaid's and Haifa's grandparents had owned farms in Palestine, but subsequently lost their domain, houses, livelihood, family histories, and identities in 1948

Gail Brussel

when the State of Israel was formed. As these families and their subsequent generations moved to Jordan and then eventually emigrated to the United States, it was the grandchildren, Zaid and Haifa, who seized back the identity of their grandparents' agricultural heritage and created the Norwich Meadows Farm.

Gail's Farm Share delivers throughout Westchester County, Manhattan, and in both Stamford and Greenwich in Connecticut. With a trucking distributor and teams of trucks and drivers, she helps the local food system expand by providing supplementary opportunities for farmers to connect to newer consumers. "My husband and I knew that we didn't have time to get out to a CSA pickup spot every week, and most of our friends simply couldn't either. And so we decided to figure out a way to create a delivery system," Gail says. This confident businesswoman saw a need and created a brand-new food business to address these local food distribution gaps. And as a trailblazer, she has generated a great deal of nationwide attention.

Zaid and Haifa's Norwich Meadows Farm has benefited tremendously from the expanded customer base provided through Farm Share. "It's gratifying to see how much support going into one farm can make a difference," Gail said. "Zaid and Haifa have been able to expand and buy more land in New Jersey. Too, Zaid can reach a wider network of consumers without having to do all of the marketing and customer service work himself. We do social media, outreach, and their website for them. Also, this farming couple works so incredibly hard. I don't know how any farmer has time to work efficiently with any of their customers because farming is so exceptionally difficult and time consuming. This is a fantastic partnership enabling me to provide customer support services for Norwich Meadows Farm and its products."

Gail invented her own job, and she absolutely loves it. And others like her are inventing never-seen-before local food businesses nationwide. Farm Share is on the cutting edge of this revolution, and with all of this continuous daily innovation, I'm eager to see what local food distribution will look like in five years. This is a brand new sector of the economy and workforce.

Farm Share Cheesy Grilled Corn

Few people get tired of eating corn in corn season, but if you do then here is a snazzy recipe that shakes up the norm for you.

SERVES 6

¼ cup whole-milk yogurt
3 tablespoons sour cream
1 teaspoon finely minced lime zest
3 tablespoons minced cilantro leaves
¾ teaspoon chili powder, divided
¼ teaspoon ground black pepper
¼ teaspoon ground cayenne pepper (optional)
3 tablespoons lime juice
½ cup grated Romano cheese
4 teaspoons olive oil
¼ teaspoon sea salt
6 large ears of corn, husks and silk removed
Sea salt and pepper to taste

Combine yogurt, sour cream, lime zest, cilantro, ¼ teaspoon chili powder, black pepper, cayenne, lime juice, and cheese in a large bowl. Stir to combine and set aside.

In a second bowl, combine oil, sea salt, and remaining chili powder. Add corn and toss to coat evenly.

Grill corn, turning occasionally, until lightly charred on all sides, about 7 to 12 minutes. Place corn in the bowl with yogurt mixture. Toss to coat evenly and serve immediately. Season to taste with sea salt and pepper.

Gail's Sweet Pomegranate Couscous

Most people won't eat pomegranates because they're a pain in the backside to peel. To make things easier, cut the pomegranate in quarters lengthwise. Over a large bowl, pull each quarter backward against the skin to invert the shape. Most of the little seeds will pop right out for you.

SERVES 6–8

2 large pomegranates
3 tablespoons freshly squeezed orange juice
¼ teaspoon black pepper
½ teaspoon orange zest
4 tablespoons maple syrup
3 tablespoons unsalted butter
1 cup fine- or medium-grain couscous
¼ cup heavy cream
1 cup whole milk
1 cinnamon stick
½ teaspoon sea salt
Sea salt and pepper to taste
½ teaspoon ground cinnamon
2 tablespoons shelled and crushed pistachios

Over a large bowl, break up the pomegranates, discarding the skin and whitish pulp and saving the seeds and juices. Toss the seeds and juices with orange juice, black pepper, orange zest, and maple syrup. Cover and let stand for at least 2 hours.

In a saucepan with a tight-fitting lid, melt 3 tablespoons of butter over medium heat. Add couscous and stir vigorously to toast in the melted butter. Off the heat, slowly add heavy cream and continue stirring. Return the pot to the heat, stir in the milk, and add the cinnamon stick and sea salt. Simmer over very low heat for about 5 minutes. Remove from the heat and cover. Allow to rest until the couscous is light and fluffy, about 10 minutes. Toss couscous with a fork to break up any clumps. Remove the cinnamon stick.

Add the pomegranate seeds and their soaking liquid to the couscous. Season to taste with sea salt and pepper. Spoon into individual dishes and serve with a dash of cinnamon and a pinch of crushed pistachios. Serve warm.

Hemlock Hill Farm

500 Croton Ave., Cortlandt, NY 10552
(914) 737-2810 | www.hemlockhillfarm.com

John and Laura De Maria make a lovely father-daughter couple. After a few photographs, Laura quietly disappears, leaving me to chat alone with John out in the pasture. "She sticks me with all of these interviews," John smiles apologetically. Actually, I don't think he minds. He has a story to tell.

As one of the oldest working farms in Westchester County, the De Maria family farm has been striving to provide the community with natural, wholesome, farm-raised products for more than seventy years. In short, Hemlock Hill Farm aspires to preserve what is left of family farming. And since John De Maria has four daughters, Laura will become the third-generation farmer in this family.

John and Laura De Maria

In 1939, Bronx native Nicholas De Maria established the 120-acre Hemlock Hill Farm. Despite a terminal illness, Nicholas refused to sell the farm to developers as the other surrounding farms did when New York's dairy industry went belly-up and the Van Cortlandt Dairy Bottling Plant went out of business. Because of his father's illness, John received an honorable discharge from the United States Army to return home to operate the family farm. Transitioning Hemlock Hill Farm from dairy and apple orchards to grass-and-brewery-grain-fed beef, pork, and chickens, Farmer John built a butcher shop and farm store that still showcase the very best of locally produced meats, vegetables, eggs, cheeses, and charcuterie.

Today, John manages 180 head of pasture-fed beef cattle. (At one point, John raised only 1,200 pigs, but the ordering preferences of his consumers changed.) He raises, slaughters, and custom butchers all of his hand-raised animals almost exclusively at the farm. This closed-loop life-and-death cycle of animal sanctity in our food supply is rare, precious, and actually, quite harmonious. Animals raised for food at Hemlock Hill Farm spend their entire life cycles side-by-side with John and Laura. They are not trucked thousands of miles across our highways. They do not spend frightening nights standing, crowded together on metal grates, at truck stops while drivers rest. They are not unloaded in strange

Hemlock Hill
FARM STORE
Fresh Meats & Produce
Local Honey
For Sale

facilities, dunked forcefully into sanitation baths, and corralled through ever-narrowing tunnels to their impersonal assembly-line deaths. Animals raised for food by John and Laura end their lives with dignity. They are surrounded by the green-grass pastures of wide-open spaces, song birds, ducks, and friendly, gentle people.

I asked John how agriculture has changed over his lifetime. "The Farm Bureau has been great," he said. "And [former New York] Governor Pataki really helped us by implementing an agricultural property tax reform to help make farming more affordable than it had been in the past. But it's a little too late, because the farmers are mostly gone, and only a few of us remain. It's a struggle for all of us." Too, Hemlock Hill Farm received New York State Farm Service Agency (FSA) moneys through the US Department of Agriculture, but during the financial crisis the granting program was temporarily suspended, and those moneys appropriated for agriculture were spent elsewhere in state budgets. It took John more than five years to receive the federal money promised to his farm. "It cost us a lot of money to survive, and we were continuously threatened with foreclosure."

The FSA program is not a grant program; it is a "pay to play" opportunity. In general, one must spend money and invest in equipment or capital projects and then the program exists to repay those moneys. John De Maria is of retirement age. He's worked extremely hard his entire career and found himself suddenly fighting foreclosure simply because the funds promised to reimburse Hemlock Hill Farm disappeared. "The last few years have been both frightening and very difficult." John pauses, surprised to find he's choking back tears. "Thank God for my daughter Laura coming in to help me. It's been a relief, and she's helped tremendously. She'll take over operations, and hopefully this farm will survive forever. That girl is my hero."

JOHN DEERE
1010

Farmer John's Cabbage Dijonnaise

Cabbage is an excellent source of vitamin C and this recipe is perfect for cold and flu season and is an easy dinner to make when you're not feeling particularly well. If you'd prefer a vegetarian version, skip the sausage altogether or substitute with tofu, tempeh, or simply add more vegetables.

SERVES 4

½ head cabbage, cut into quarters through the core
2 tablespoons olive oil
½ cup sliced red onion
2 large sausages of your choice, casings removed, and crumbled, about 2 cups
¼ cup white wine
2 tablespoons vodka or cider vinegar
1 teaspoon maple syrup (if using vodka)
1 cup chicken stock
1 tablespoon Dijon mustard
½ teaspoon red pepper flakes
1 tablespoon fresh thyme leaves
Sea salt and pepper to taste
¼ cup chopped fresh parsley leaves
1 tablespoon unsalted butter, sliced

Bring a large pot of water to a boil. Add cabbage pieces and boil for 5 to 7 minutes until the cabbage is crisp-tender. Drain well. Cut off the core and shred the leaves into bite-sized pieces.

Heat oil in a large skillet over medium-high heat. Add red onions and sauté until beginning to brown. Add sausage and sauté for another 2 minutes or until the meat starts to brown. Add cabbage and stir to combine. Pour in the wine, vodka, maple syrup, and chicken stock. Stir in the mustard, red pepper flakes, and thyme.

Bring to a boil, cover, and simmer for 20 minutes or until the cabbage is very soft. Season to taste with sea salt and pepper. Sprinkle with parsley, add butter, and serve warm.

Hemlock Hill Beef Stock

Typically, beef stock is a labor- and energy-intensive process requiring long hours to roast beef bones. This version creates a very flavorful stock using beef shanks. The trick to great flavor for any stock is the slow, low heat simmering. You don't need to buy commercial beef stock again.

MAKES 2 QUARTS

2 tablespoons grapeseed oil, plus additional if necessary
6 pounds beef shanks, meat removed and cut into large chunks, bones reserved
1 large onion, halved
4 large carrots, peeled and quartered crosswise
2 quarts boiling water
½ cup whiskey or bourbon
½ teaspoon sea salt
1 tablespoon maple syrup
Sea salt and pepper to taste

In a large stockpot or dutch oven, heat the grapeseed oil over medium-high heat. Brown the meat, bones, onion, and carrots in batches, being careful not to overcrowd the pan and adding additional oil as needed. Remove seared contents to a large bowl and set aside.

Add ½ cup water to the hot stockpot and bring to a rapid boil. Add the whiskey and boil, scraping any browned bits off the bottom of the pan, until reduced by half and syrupy.

Return the carrots and onion to the pot and top with beef meat and bones. Reduce the heat to low and cover, allowing the meat and bones to release about ¾ cup of liquid, about 20 minutes.

Increase the heat to medium-high and add the remaining boiling water and sea salt. Bring to a simmer, reduce the heat to low again, partially cover, and barely simmer until the meat is tender, about 1½ to 2 hours.

Strain the stock into a container to cool. Add maple syrup and season to taste with sea salt and pepper. Discard the bones, onion, and carrots. Reserve the meat for use later. When cool, skim the fat from the top and discard.

Use immediately or store in the refrigerator for up to 5 days.

FARM-AT-HOME GARDEN
WELCOME
Farm-at-Home Garden
Rules agreed to upon entering Garden
Please Keep Shut to exclude Hens

Hilltop Hanover Farm & Environmental Center

1271 Hanover St., Yorktown Heights, NY 10598
(914) 962-2368 | www.hilltophanoverfarm.org

There is a very special place in Yorktown Heights, New York. When you're there, if you pivot 360-degrees, you will see spectacular views that make you forget you're standing in suburban Westchester County and think instead you are deep in the heart of Vermont's rolling hills. The skyline is layered with the blue and purple shades of hills on the horizon. Hawks and eagles fly overhead, and the wind tickles across the hilltop in a lazy, noncommittal fashion. Which would you choose: a sixty-unit condominium complex or a nonprofit farm preserve with green fields as far as the eye can see? Luckily, Westchester County has currently chosen the latter.

To help preserve the Croton Watershed, the county purchased 187 of the original three hundred acres of Hilltop Hanover Farm for open-space preservation and conservation of their agricultural heritage. In 2010, the Friends of Hilltop Hanover Farm & Environmental Center Inc. was formed to assist the county in raising funds and help with the day-to-day operations of the facility. One of the goals of this organization is eventually to generate enough financial support and create self-sustainable programming to take over the financial burdens of the farm's operations, thereby helping Westchester County reduce the strain on its budget that this amazing community gift of preserved open space has created. With the current economic crisis, the county was forced to severely cut the operating budget of the Hilltop Hanover working farm. As a consequence, the Friends group is rallying community members, granting organizations, and large donors to help fill this monetary gap and stave off any future development of this valuable property. "It's a little difficult to be underfunded and understaffed," the manager, Lucille Munz, said. "But everyone is in the same boat, and we all just have to pull together."

Lucille Munz

Hilltop Hanover Farm & Environmental Center is a working farm and regional environmental education center that provides opportunities for practicing healthy and sustainable food production, for learning small-scale suburban and urban farming techniques, and for engaging with sustainable living practices. Operated by a tremendously dedicated board of directors and managed by Lucille, this working farm offers demonstration models for backyard farming, rainwater harvesting, composting, and green-roof technology. Visitors can hike the farm's three and a half miles of wooded trails, picnic on the farm grounds, attend numerous classes, purchase CSA shares, or participate in the farm's U-pick produce program throughout the growing season.

The Hilltop Hanover Farm is the only completely U-pick vegetable farm in Westchester County. The resident farmer, Brett Alcaro, tends five acres of vegetables and flowers. People from within a hundred-mile radius—Connecticut, New York, and New Jersey—come to harvest their own vegetables. This is a vitally important service, and one that is received with high praise. "You can pick your own tomatoes, and you can dig your own carrots," Lucille says. "That directly connects people with the land, and they start to see the importance of a farm." Coming here becomes an experience and more than just

a destination parking lot to procure pre-picked and sanitized vegetables. Food remaining on the vine or in the soil is harvested by volunteers and brought to food pantries.

Lucille believes that everyone can farm at home, whether you have acres of open backyard space or a simple apartment with sun-drenched windowsills. Under her watchful stewardship, this suburban farm preserve demonstrates that learning to actually feed ourselves is a fundamental step toward mitigating many of the health and economic difficulties we face as individuals and communities.

It's difficult not to giggle when Lucille starts talking about sedums and succulents. She was so stoic and serious in her crisp white shirt as she discussed the valuable benefits that these green-roof plants have on the quality of groundwater in the Croton Watershed. These are serious topics indeed, and I don't know why I started to giggle, but I did. After a minute or so, while I continuously tried (and failed) to compose myself, Lucille broke into a wide grin. Her dry, quick-fire wit is razor sharp. The words *sedums* and *succulents* are great comedy fodder for someone as smart as Lucille. I spent the rest of the afternoon with Lucille, munching on Sun Gold tomatoes pulled fresh from the vine and roaring with laughter.

Lucille's Watercress Soup with Blue Cheese Croutons

> *Worries go down better with soup than without.*
> —*Jewish Proverb*

SERVES 6-8

1 baguette
2 tablespoons olive oil
1 tablespoon unsalted butter
3 large shallots, sliced
1 cup peeled and diced russet potatoes
1 cup peeled and diced parsnips
3 cups peeled, cored, and diced ripe Bartlett pears
4 cups stock of your choice (see p. ix)
3 packed cups fresh watercress with stems, plus extra for garnish
1 cup whipping cream, divided
Sea salt to taste
½ cup crumbled stilton or other blue cheese

Preheat oven to 350°F. Cut 6 ½-inch-thick slices of baguette on the diagonal to make large croutons. Brush both sides of bread slices with olive oil and lay on a baking sheet. Bake, turning slices half way through, for 6 to 8 minutes or until toasted and hard. Set aside but keep warm.

Heat butter in a large stockpot over medium heat until bubbling and just beginning to brown. Add shallots and sauté for 5 minutes until softened. Add potatoes, parsnips, pears, and stock. Bring to a boil, cover and simmer for 25 minutes or until the potatoes and parsnips are tender and cooked through.

Add all of the watercress to the hot soup on medium heat. Stir constantly until the watercress is wilted, about 4 or 5 minutes. Immediately puree the soup using a blender or an immersion blender until completely smooth.

Return the soup to a clean stockpot and add ½ cup cream. Heat until hot and season to taste with sea salt. Remove from heat and set aside.

Top croutons with stilton cheese and press lightly. Heat in the oven for 3 or 4 minutes to soften the cheese.

To serve, ladle soup into warmed serving bowls. Drizzle remaining whipping cream around the perimeter of the soup bowls to garnish. Float a melted stilton crouton in the center of each bowl and garnish with a fresh watercress sprig. Serve warm.

Hilltop Hanover Enchiladas with Tomatillo Sauce

Here's a small secret about me: I actually dislike cooked tomatoes. So in my house, we use everything else. Try tomatillos. If you don't have those, you can totally bastardize this recipe by using 6 medium red apples. Yum!

SERVES 4

1 garlic clove, peeled
1 jalapeño pepper, stemmed and quartered
12 medium-size tomatillos, husked, rinsed, and quartered
¾ cup loosely packed cilantro leaves, plus additional for garnish
3 pieces thick-cut bacon
½ pound ground beef
Sea salt and pepper to taste
2 cups stock of your choice (see p. ix)
3 tablespoons heavy cream or sour cream
½ teaspoon maple syrup
8 ounces oyster or button mushrooms, stemmed and sliced
1 teaspoon olive oil (optional)
1 large red onion, thinly sliced
7 cups fresh spinach leaves
12 corn tortillas
1 cup crumbled queso fresco or feta cheese

Preheat oven to 350°F. With a food processor running, drop in garlic and jalapeño piece by piece. Add the tomatillos and the cilantro in batches, processing until smooth.

In a large sauté pan, cook bacon until crispy. Remove bacon and crumble; set aside. Reserve half the bacon drippings and place in a second large sauté pan. Using the bacon grease in the second pan, thoroughly cook all of the ground beef. Salt and pepper to taste, drain the ground beef on paper towels in a small bowl, and set aside. Reserve this second sauté pan to cook the vegetables later.

In the first sauté pan used to cook the bacon, add the garlic-tomatillo puree and cook over medium-high heat, stirring constantly, until the mixture has reduced to the consistency of thick tomato sauce, about 10 minutes. Add the stock and simmer for about 10 minutes to blend the flavors. Add the cream and maple syrup and bring to a simmer. Season to taste with sea salt and pepper. Reduce heat to low, cover, and keep warm.

While the sauce is simmering, make the filling in the second sauté pan used for cooking the ground beef. Using the fat and flavors already in this sauté pan, add the

mushrooms and cook, stirring constantly, over high heat, until the mushrooms begin to brown. Add a teaspoon of olive oil if you need to. Add three-fourths of the sliced red onions, reserving the remaining for garnish, and sauté, stirring constantly, until the onions are translucent. Add the spinach and ground beef and sauté until the spinach is wilted. Season to taste with sea salt and pepper. Remove from heat and cover to keep warm.

Holding tortillas by an edge, dredge each through the tomatillo sauce and lay it flat on a plate. Spoon ¼ cup filling down the center of the tortilla, roll it up, and lay it seam side down on a plate. Repeat for all the tortillas, serving 3 enchiladas per plate.

In the meantime, bring the tomatillo sauce back to a near boil. Spoon one-fourth of the tomatillo sauce over each plate, sprinkle with a quarter of the crumbled cheese, and garnish with reserved onion and cilantro sprigs. Serve immediately.

Ironbound Farm

1176 Williams Dr., Shrub Oak, NY 10579
(845) 582-2565 | www.ironboundfarm.com

Ironbound Farm was created in 2006 on five acres of Nancy McDaid's family land in Putnam Valley. Despite its small size, this little homestead produces Nigerian dwarf goats, Alpine goats, free-range chickens and eggs, honey, vegetables, herbs, and Nancy's homemade preserves. They practice "square-foot gardening" organic vegetable methods, which utilize raised beds and specific quantities of seeds planted for a specific square footage of soil. You can find everything here from rhubarb to purple pod beans, tat choi, edamame, kohlrabi, and sorrel. Besides the farm animals, Nancy tends four children and a collie, a coonhound, and a Great Dane.

Nancy McDaid

Nancy McDaid started slowly building Ironbound Farm so that she could produce food for her family. "I just wanted to know where our food was coming from," she asserted. "I wanted food that wasn't full of chemicals." In addition to all of her frantic motherhood chores with four kids (including twins), she added more work to her plate by building a farm. "It's a substitute for having more children!" she laughs. "Now that they're in school, this gives me something meaningful to do that provides nourishment for them when they return home on the school bus."

Cornish cross is the breed of chicken that we're most familiar with. They're what we buy, plucked and cleaned, from the grocery store. The agribusiness-brand Cornish cross chickens are genetically bred to grow extremely large very fast, making them ready for market faster than a local farmer's meat birds. A well-documented fact, these chickens are force fed, growing so fast that often their legs cannot hold them up and occasionally break, forcing these chickens to spend their lives lying in dirt and feces and requiring antibiotics to prevent the spread of festering sores, diseases, or respiratory failure. Commercially raised chickens are often bald, with large patches of exposed skin, since their bodies grow faster than their feathers can sprout. These deficiencies

make them more susceptible to extreme weather conditions and injuries from other chickens higher up on the established pecking order. For producers primarily concerned with profit, more chickens can be raised throughout the year than typical breeds, and more such chickens can be raised together in darkened steel feedlots.

At first, Nancy attempted to raise these Cornish cross chickens at Ironbound Farm. "For some reason they didn't know how to eat the greens or search around and scratch at the ground for bugs. They could hardly walk!" Nancy said. After that, she decided to raise a heritage breed called colored rangers. "They are such a pleasure to have around, and they act like real chickens! They scratch for bugs, eat the weeds, and have a good time stomping through the place." Such has been the journey and life lessons of Mr. and Mrs.

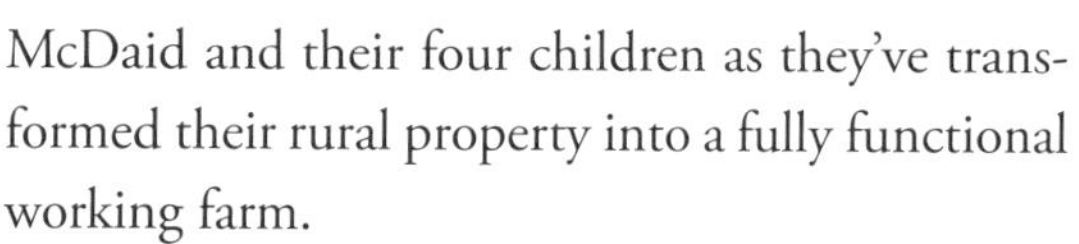

McDaid and their four children as they've transformed their rural property into a fully functional working farm.

Ironbound Farm is not only a food-producing paradise, it's also a fun and phenomenally interesting place for kids to grow up. The eldest daughter is thirteen years old, the twin boys are ten, and the little guy is eight. They can milk the goats, collect eggs, hike the back forest trails, pick apples or pears or peaches, or play soccer with each other in the cleared, grassy fields. The boys prefer garden chores while the lone female sibling nurtures all of the animals. Growing up surrounded with the freshest food possible, these kids are incredibly label savvy. "They'll read labels and play games quizzing each other over the breakfast table in the morning. I think that's hilarious," Nancy said.

Many people believe that to grow food you must be an avid full-time farmer, quit your job, and live off the land. "Not so," says Nancy while leaning comfortably against Schmidgeon, the family's new milk cow. "I am always encouraging people to just start producing food in their backyard, even if it's not perfect. Farming certainly works well for me and my entire family."

Nancy's Limey Coconut Shrimp

Here's a random fact about coconut that you may not know: Inside a coconut is a cavity filled with coconut water, which is sterile until opened. It mixes easily with blood, and was used during World War II in emergency transfusions. It can also serve as an emergency short-term intravenous hydration fluid.

SERVES 4

1 pound large, tail-on shrimp with tails
2 cups panko bread crumbs
½ cup shredded unsweetened coconut
2 tablespoons finely minced lime zest
½ cup all-purpose flour
¼ teaspoon sea salt
2 eggs
¼ cup unsweetened coconut milk
Juice of 1 lime
2 cups peanut oil or grapeseed oil
1 lime, cut into wedges
Sea salt and pepper to taste

Preheat oven to 175°F. Place a baking sheet lined with several paper towels in the preheated oven.

Peel shrimp, if necessary, leaving tails attached. Thoroughly rinse and dry the shrimp. Set aside.

Combine panko bread crumbs, coconut, and lime zest in one bowl and set aside. In a second bowl, combine flour and sea salt. In a third small bowl, whisk eggs with coconut milk and the lime juice until well mixed.

Cover a baking sheet with waxed paper. Holding it by the tail, dip a shrimp into flour, tapping off any excess that does not adhere. Dip the floured shrimp into the egg mixture to coat lightly and then dip the shrimp into the panko and coconut, pressing so that the mixture adheres. Place on waxed paper and proceed to coat the remaining shrimp.

Heat a few inches of oil in a deep, heavy frying pan over medium-high heat until hot. Fry several shrimp at a time for 3 to 4 minutes, turning often, or until golden and crispy. Drain on paper towel-lined baking sheet and keep them warm in the oven while frying the remaining shrimp.

Serve immediately with lime wedges or with Camp Hill Goat Cheese Fettuccine (see p. 164). Season to taste with sea salt and pepper.

Ironbound Farm Orange Salad with Almonds

My favorite thing is to make this salad for lunch and then to combine the leftovers with roasted brussels sprouts for dinner. Whenever I do, I always think of P. J. O'Rourke, who said, "A fruit is a vegetable with looks and money. Plus, if you let fruit rot, it turns into wine, something brussels sprouts never do."

SERVES 4

4 teaspoons golden raisins
3 tablespoons olive oil
2 tablespoons raspberry or apple cider vinegar
¼ teaspoon honey
½ teaspoon sea salt
½ teaspoon pepper
4 oranges, peeled, seeded, each section sliced in half
½ red onion, sliced paper thin
4 teaspoons toasted almond slivers
1 tablespoon minced fresh mint leaves

Soak the raisins in warm water to cover for 15 minutes. Drain and set aside.

In a small bowl whisk together olive oil, raspberry vinegar, honey, sea salt, and pepper to make a vinaigrette. Correct seasoning, if necessary.

Arrange the orange slices on individual salad places. Scatter with onion slices, almonds, and raisins and pour the dressing over the orange slices. Garnish with mint leaves. Serve at room temperature.

Appendix

Explore All Things Local in New York City

Citizens Committee for New York City
www.citizensnyc.org

City Farming NYC Meetup
www.meetup.com/City-Farming-NYC

The Entrepreneur Space
www.queensny.org/qedc/entrepreneur_space

Farmers Market Federation of New York
www.nyfarmersmarket.com

Food NYC Report
www.mbpo.org/uploads/policy_reports/mbp/FoodNYC.pdf

Food Rights Network
www.foodrightsnetwork.org

Foods of New York Tours
www.foodsofny.com

Glynwood Center
www.glynwood.org

Glynwood Center Keep Farming Program
www.glynwood.org/programs/keep-farming

Grow NYC
www.grownyc.org

Just Food
www.justfood.org

Northeast Organic Farming Association of New York
www.nofany.org

NYC Parks GreenThumb
www.greenthumbnyc.org

Slow Food NYC
www.slowfoodnyc.org

Recipe Index

WELCOME

General Index

About the Author

Emily Brooks is one of the principal authorities on the local food and sustainable agriculture movements. Founder of Edibles Advocate Alliance (www.ediblesadvocatealliance.org) and the founder of BRIDGES Healthy Cooking School, Chef Emily nurtures social entrepreneurs who support local agriculture, sustainable farming, food sovereignty, and sustainable food systems. She is also the author of *Connecticut Farmer & Feast* and lives in Connecticut, with her dog, Lady.

MW00764510
NEW YORK CITY
FARMER & FEAST